Problems and Solutions for Undergraduate Real Analysis I

by Kit-Wing Yu, PhD

kitwing@hotmail.com

ISBN: 978-988-78797-4-9 (eBook)
ISBN: 978-988-78797-5-6 (Paperback)

About the author

Dr. Kit-Wing Yu received his B.Sc. (1st Hons), M.Phil. and Ph.D. degrees in Math. at the HKUST, PGDE (Mathematics) at the CUHK. After his graduation, he has joined United Christian College to serve as a mathematics teacher for at least seventeen years. He has also taken the responsibility of the mathematics panel since 2002. Furthermore, he was appointed as a part-time tutor (2002 – 2005) and then a part-time course coordinator (2006 – 2010) of the Department of Mathematics at the OUHK.

Apart from teaching, Dr. Yu has been appointed to be a marker of the HKAL Pure Mathematics and HKDSE Mathematics (Core Part) for over thirteen years. Between 2012 and 2014, Dr. Yu was invited to be a Judge Member by the World Olympic Mathematics Competition (China). In the area of academic publication, he is the author of two books "*A Complete Solution Guide to Principles of Mathematical Analysis*" and "*Mock Tests for the ACT Mathematics*". Besides, he has published over twelve research papers in international mathematical journals, including some well-known journals such as J. Reine Angew. Math., Proc. Roy. Soc. Edinburgh Sect. A and Kodai Math. J.. His research interests are inequalities, special functions and Nevanlinna's value distribution theory.

Preface

The aim of *Problems and Solutions for Undergraduate Real Analysis I*, as the name reveals, is to assist undergraduate students or first-year students who study mathematics in learning their first rigorous real analysis course.

"The only way to learn mathematics is to do mathematics." – Paul Halmos. My learning and teaching experience has convinced me that this assertion is definitely true. In fact, I believe that "doing mathematics" means a lot to everyone who studies or teaches mathematics. It is not only a way of writing a solution to a mathematical problem, but also a mean of reflecting mathematics deeply, exercising mathematical techniques expertly, exchanging mathematical thoughts with others effectively and searching new mathematical ideas unexpectedly. Thus I hope everyone who is reading this book can experience and acquire the above benefits eventually.

The wide variety of problems, which are of varying difficulty, include the following topics: Elementary Set Algebra, the Real Number System, Countable and Uncountable Sets, Elementary Topology on Metric Spaces, Sequences in Metric Spaces, Series of Numbers, Limits and Continuity of Functions, Differentiation and the Riemann-Stieltjes Integral. Furthermore, the main features of this book are listed as follows:

- The book contains 230 problems, which cover the topics mentioned above, with *detailed* and *complete* solutions. As a matter of fact, my solutions show every detail, every step and every theorem that I applied.

- Each chapter starts with a brief and concise note of introducing the notations, terminologies, basic mathematical concepts or important/famous/frequently used theorems (without proofs) relevant to the topic.

- Three levels of difficulty have been assigned to problems:

Symbol	Level of difficulty	Meaning
⋆	Introductory	These problems are basic and every student must be familiar with them.
⋆ ⋆	Intermediate	The depth and the complexity of the problems increase. Students who targets for higher grades must study them.
⋆ ⋆ ⋆	Advanced	These problems are very difficult and they may need some specific skills.

- Different colors are used frequently in order to highlight or explain problems, examples, remarks, main points/formulas involved, or show the steps of manipulation in some complicated proofs. (ebook only)

- An appendix about mathematical logic is included. It tells students what concepts of logic (e.g. techniques of proofs) are necessary in advanced mathematics. If you are familiar with these, you may skip it. Otherwise, you are strongly recommended to spend time to read at least §A.3 to §A.5.

If you find any typos or mistakes, please feel free to send your valuable comments or opinions to

kitwing@hotmail.com

Any updated errata of this book or news about my new book will be posted on my new website:

https://sites.google.com/view/yukitwing/

Kit Wing Yu
September 2018

List of Tables

4.1 Properties of the interior and the closure of E in X. 29

A.1 The truth table of $p \wedge q$. 188

A.2 The truth table of $p \vee q$. 188

A.3 The truth table of $\sim p$. 189

A.4 The truth table of $p \to q$. 189

A.5 The truth table of $p \to q$. 189

A.6 The truth table of $p \to q$ and $\sim q \to \sim p$. 190

A.7 The tautology $(p \vee q) \vee (\sim p)$ 190

A.8 The contradiction $(p \wedge q) \wedge (\sim p)$ 190

Contents

Preface   **v**

List of Tables   **vii**

1   Elementary Set Algebra   **1**

    1.1   Fundamental Concepts   1

    1.2   Sets, Functions and Relations   4

    1.3   Mathematical Induction   6

2   The Real Number System   **9**

    2.1   Fundamental Concepts   9

    2.2   Rational and Irrational Numbers   10

    2.3   Absolute Values   12

    2.4   The Completeness Axiom   13

3   Countable and Uncountable Sets   **19**

    3.1   Fundamental Concepts   19

    3.2   Problems on Countable and Uncountable Sets   20

4   Elementary Topology on Metric Spaces   **27**

    4.1   Fundamental Concepts   27

    4.2   Open Sets and Closed Sets   31

    4.3   Compact Sets   38

    4.4   The Heine-Borel Theorem   44

    4.5   Connected Sets   45

5   Sequences in Metric Spaces   **49**

    5.1   Fundamental Concepts   49

    5.2   Convergence of Sequences   53

    5.3   Upper and Lower Limits   59

5.4   Cauchy Sequences and Complete Metric Spaces   65

5.5   Recurrence Relations   70

6 Series of Numbers   75

6.1   Fundamental Concepts   75

6.2   Convergence of Series of Nonnegative Terms   79

6.3   Alternating Series and Absolute Convergence   87

6.4   The Series $\Sigma_{n=1}^{\infty} a_n b_n$ and Multiplication of Series   90

6.5   Power Series   94

7 Limits and Continuity of Functions   97

7.1   Fundamental Concepts   97

7.2   Limits of Functions 103

7.3   Continuity and Uniform Continuity of Functions 108

7.4   The Extreme Value Theorem and the Intermediate Value Theorem 116

7.5   Discontinuity of Functions 120

7.6   Monotonic Functions 122

8 Differentiation   127

8.1   Fundamental Concepts 127

8.2   Properties of Derivatives 132

8.3   The Mean Value Theorem for Derivatives 138

8.4   L'Hôspital's Rule 146

8.5   Higher Order Derivatives and Taylor's Theorem 149

8.6   Convexity and Derivatives 153

9 The Riemann-Stieltjes Integral   157

9.1   Fundamental Concepts 157

9.2   Integrability of Real Functions 162

9.3   Applications of Integration Theorems 170

9.4   The Mean Value Theorems for Integrals 182

Appendix   187

A Language of Mathematics   187

A.1   Fundamental Concepts 187

A.2   Statements and Logical Connectives 188

A.3   Quantifiers and their Basic Properties 190

A.4   Necessity and Sufficiency 191

Contents

A.5 Techniques of Proofs 192

Index   **195**

Bibliography   **199**

CHAPTER 1

Elementary Set Algebra

▌ 1.1 Fundamental Concepts

In this section, we briefly record some basic properties of sets, functions, equivalence relations and order relations. For detailed introduction to these topics, the reader can read [10, §1 - §4, pp. 4 - 36], [15, Chap. 3] and [16, §1.2 - §1.4].

▌ 1.1.1 Sets

Capital letters $E, F, \ldots$ are usually used to represent **sets** and lowercase letters $x, y, \ldots$ refer to elements of sets. The notation $x \in E$ means that x belongs to the set E. Similarly, the notation $x \notin E$ means that x does not belong to E.

Several sets are common in analysis. They are $\mathbb{N}, \mathbb{Z}, \mathbb{Q}$ and $\mathbb{R}$ which are called positive integers, integers, rational numbers and real numbers respectively. Furthermore, the set $\mathbb{R} \setminus \mathbb{Q}$ is the set of all irrational numbers. If a "+"sign (resp. "−" sign) is put in the superscript of each of the above set (except $\mathbb{N}$), then the new set takes only the positive (resp. negative) part of the base set. For instance, $\mathbb{R}^+$ means the set of all positive real numbers.

▌ 1.1.2 Basic Operations with Sets

Two sets E and F are said to be **equal**, namely $E = F$, if they consist of precisely the same elements. A set E is called a **subset** of F, in symbols $E \subseteq F$, if every element of E is an element of F. If $E \subseteq F$ but $E \neq F$, then E is called a **proper subset** of F. The set without any elements is called the **empty set** which is denoted by $\varnothing$.

Suppose that E and F are subsets of a set S. Then we define their **union**, their **intersection** and their **difference** as follows:

- **Union:** $E \cup F = \{x \in S \mid x \in E \text{ or } x \in F\}$.

- **Intersection:** $E \cap F = \{x \in S \mid x \in E \text{ and } x \in F\}$.

- **Difference:** $E \setminus F = \{x \in S \mid x \in E \text{ and } x \notin F\}$.

1

In particular, we call

$$E^c = S \setminus E$$

the **complement** of E. The following theorem states some basic operations of union, intersection and difference of sets:

Theorem 1.1. *Suppose that A, B and C are sets. Then we have*

- $(A \cup B) \cap C = (A \cap C) \cup (B \cap C)$;

- $(A \cap B) \cup C = (A \cup C) \cap (B \cup C)$;

- $(A \cup B) \setminus C = (A \setminus C) \cup (B \setminus C)$;

- $(A \cap B) \setminus C = (A \setminus C) \cap (B \setminus C)$.

1.1.3 Functions

Given two sets X and Y, we define their **Cartesian product** $X \times Y$ to be the set of all **ordered pairs** (x, y), where $x \in X$ and $y \in Y$. Formally, we have

$$X \times Y = \{(x, y) \,|\, x \in X \text{ and } y \in Y\}.$$

By a **function** (or **mapping**) f from X to Y, in symbols $f : X \to Y$, we mean that a specific "rule" that assigns to each element x of the set X a unique element y in Y. More precisely,

Definition 1.2. *[10, p. 15] A **rule of assignment** is a subset r of the Cartesian product $X \times Y$ such that each element of X appears as the first coordinate of at most one ordered pair in r.*

Now the element y is called the **value** of f at x and it is expressed as $y = f(x)$. The set X is called the **domain** of f and the set

$$f(X) = \{f(x) \in Y \,|\, x \in X\}$$

is called the **range** of f. A function $f : X \to Y$ is called **injective** (or **one-to-one**) if $x_1 \neq x_2$ implies that $f(x_1) \neq f(x_2)$. Besides, f is said to be **surjective** (or **onto**) if $f(X) = Y$. In particular, a function f is **bijective** if it is both injective and surjective.

If $E \subseteq Y$, then we define

$$f^{-1}(E) = \{x \in X \,|\, f(x) \in E\}.$$

We call $f^{-1}(E)$ the **inverse image** of E under f. In particular, if $E = \{y\}$, then we have

$$f^{-1}(y) = \{x \in X \,|\, f(x) = y\}.$$

Definition 1.3. *Given functions $f : X \to Y$ and $g : Y \to X$. Their **composition** $g \circ f$ is the function $g \circ f : X \to Z$ defined by*

$$(g \circ f)(x) = g(f(x))$$

for each $x \in X$.

▌1.1.4 Equivalence Relations

Definition 1.4. *Given a set* X. *By a* **relation** *on a set* X, *we mean a subset* $\mathcal{R}$ *of the Cartesian product* $X \times X$. *If* $(x, y) \in \mathcal{R}$, *then* x *is said to be in the relation* $\mathcal{R}$ *with* y. *We use the notation* $x\mathcal{R}y$ *to mean this situation.*

One of the most interesting and important relations is the **equivalence relation**. In fact, a relation $\mathcal{R}$ on a set X is said to be an **equivalence relation** if it satisfies the following three conditions:

- **(1) Reflexivity:** $x\mathcal{R}x$ for every $x \in X$.

- **(2) Symmetry:** If $x\mathcal{R}y$, then $y\mathcal{R}x$.

- **(3) Transitivity:** If $x\mathcal{R}y$ and $y\mathcal{R}z$, then $x\mathcal{R}z$.

To make things simple, we will use the notation "$\sim$" to replace the letter "$\mathcal{R}$" in an equivalence relation.

Given an equivalence relation $\sim$ on a set X. Pick $x \in X$. We define a subset $E \subseteq X$, called the **equivalence class** determined by x, by

$$E_x = \{y \in X \mid y \sim x\}.$$

A basic fact about equivalence classes is the following result:

Theorem 1.5. *Two equivalence classes* E_x *and* E_y *are either disjoint or coincide.*

By this concept, we know that the family of equivalence classes form a **partition** of the set X.

▌1.1.5 Order Relations

Besides equivalence relations, **order relations** are another important types of relations. In fact, a relation "$<$" (means "less than") is called an order relation if it satisfies the following properties:

- **(1) Comparability:** If $x \neq y$, then either $x < y$ or $y < x$.

- **(2) Nonreflexivity:** If $x < y$, then $x \neq y$.

- **(3) Transitivity:** If $x < y$ and $y < z$, then $x < z$.

With the order relation "$<$", concepts of inequality, interval, (upper and lower) bound, supremum and infimum of real numbers can be developed.

1.2 Sets, Functions and Relations

Problem 1.1

($\star$) *Suppose that E and F are subsets of a set S. Let E^c be the complement of E in S. Prove De Morgan's laws: $(E \cap F)^c = E^c \cup F^c$ and $(E \cup F)^c = E^c \cap F^c$.*

Proof. Suppose that $x \in (E \cap F)^c$. Then it must be the case that $x \notin E$ or $x \notin F$ which mean $x \in E^c$ and $x \in F^c$, i.e.,

$$(E \cap F)^c \subseteq E^c \cup F^c. \tag{1.1}$$

Conversely, if $x \in E^c \cup F^c$, then we have $x \notin E$ or $x \notin F$. Thus we have $x \notin E \cap F$, i.e., $x \in (E \cap F)^c$ and then

$$E^c \cup F^c \subseteq (E \cap F)^c. \tag{1.2}$$

Hence the first identity follows from the set relations (1.1) and (1.2).

Since $(E^c)^c = E$, the second identity follows immediate from the first identity if we replace E and F by E^c and F^c respectively. This finishes the proof of the problem. ∎

Problem 1.2

($\star$) *Suppose that E and F are subsets of a set S. Prove that $E \setminus F = E \cap F^c$.*

Proof. Suppose that $x \in E \setminus F$. Then it is equivalent to the statement $x \in E$ and $x \notin F$ or $x \in E$ and $x \in F^c$. Finally, the last statement is equivalent to $x \in E \cap F^c$. This proves

$$E \setminus F \subseteq E \cap F^c. \tag{1.3}$$

Conversely, let $x \in E \cap F^c$. Then we have $x \in E$ and $x \in F^c$, so $x \in E$ and $x \notin F$. Therefore, it means that $x \in E \setminus F$, i.e.,

$$E \cap F^c \subseteq E \setminus F. \tag{1.4}$$

Now the identity follows from the set relations (1.3) and (1.4). This completes the proof of the problem. ∎

Problem 1.3

($\star$) *Let E and F be subsets of S. Prove that $(E \setminus F) \cup F = E$ if and only if $F \subseteq E$.*

Proof. By Problem 1.2, we know from the fact $S = F \cup F^c$ that

$$(E \setminus F) \cup F = (E \cap F^c) \cup F = (E \cup F) \cap (F^c \cup F) = (E \cup F) \cap S = E \cup F.$$

Thus $(E \setminus F) \cup F = E$ if and only if $E = E \cup F$ if and only if $F \subseteq E$. This completes the proof of the problem. ∎

Problem 1.4

($\star$) Let I, X and Y be sets. Let $f : X \to Y$ be a function and $E_i \subseteq X$ for $i \in I$. Show that

$$f\left(\bigcup_{i \in I} E_i\right) = \bigcup_{i \in I} f(E_i) \quad \text{and} \quad f\left(\bigcap_{i \in I} E_i\right) \subseteq \bigcap_{i \in I} f(E_i). \tag{1.5}$$

Show, by an example, that the inclusion of the set relation in (1.5) can be proper.

Proof. Let $y \in f\left(\bigcup_{i \in I} E_i\right)$. Then there exists $x \in \bigcup_{i \in I} E_i$ such that $f(x) = y$. In other words, there exists $x \in E_i$ such that $f(x) = y$ *for some* $i \in I$ which imply that $y \in f(E_i)$ for some $i \in I$, i.e., $y \in \bigcup_{i \in I} f(E_i)$ or

$$f\left(\bigcup_{i \in I} E_i\right) \subseteq \bigcup_{i \in I} f(E_i). \tag{1.6}$$

Conversely, if $y \in \bigcup_{i \in I} f(E_i)$, then $y \in f(E_i)$ *for some* $i \in I$ so that $y = f(x)$ for some $x \in E_i \subseteq \bigcup_{i \in I} E_i$. Thus we have $y \in f\left(\bigcup_{i \in I} E_i\right)$ which means

$$\bigcup_{i \in I} f(E_i) \subseteq f\left(\bigcup_{i \in I} E_i\right). \tag{1.7}$$

Now the first assertion follows from the set relations (1.6) and (1.7).

For the second assertion, since we always have $\bigcap_{i \in I} E_i \subseteq E_i$ *for every* $i \in I$, we must have $f\left(\bigcap_{i \in I} E_i\right) \subseteq f(E_i)$ for every $i \in I$ and this proves that

$$f\left(\bigcap_{i \in I} E_i\right) \subseteq \bigcap_{i \in I} f(E_i).$$

To show that the inclusion can be proper, let us consider $E_1 = \{1\}, E_2 = \{2\}, X = \{1,2\}$ and $Y = \mathbb{R}$. Suppose that $f : X \to \mathbb{R}$ is given by $f(x) = 1$ for $x \in X$. Since $f(E_1) = f(E_2) = \{1\}$, we have $f(E_1) \cap f(E_2) = \{1\}$. However, we know that $E_1 \cap E_2 = \varnothing$ which means that $f(E_1 \cap E_2) = \varnothing$. In this case, we have[a]

$$f(E_1 \cap E_2) \subset f(E_1) \cap f(E_2).$$

This ends the proof of the problem. ∎

Problem 1.5

($\star$) Let X and Y be sets. Let $f : X \to Y$ be a function and $E \subseteq X$. Prove that

$$f^{-1}(E^c) = [f^{-1}(E)]^c. \tag{1.8}$$

---

[a]Here we use the fact that the empty set is a subset of every set, see [13, Chap. 2, Exercise 1, p. 43].

Proof. Let $x \in f^{-1}(E^c)$. Then $f(x) \in E^c$ or $f(x) \notin E^c$. In other words, we have $x \notin f^{-1}(E)$ and so $x \in [f^{-1}(E)]^c$, i.e.,

$$f^{-1}(E^c) \subseteq [f^{-1}(E)]^c. \tag{1.9}$$

Conversely, if $x \in [f^{-1}(E)]^c$, then $x \notin f^{-1}(E)$ so that $f(x) \notin E$. Therefore, we have $f(x) \in E^c$ and this implies that $x \in f^{-1}(E^c)$, i.e.,

$$[f^{-1}(E)]^c \subseteq f^{-1}(E^c). \tag{1.10}$$

Combining the set relations (1.9) and (1.10), we have the desired result (1.8), completing the proof of the problem. ∎

Problem 1.6

$(\star)$ *Suppose that $f : X \to Y$, $g : Y \to Z$ and $E \subseteq Z$. Prove that*

$$(g \circ f)^{-1}(E) = f^{-1}(g^{-1}(E)).$$

Proof. We note that $x \in (g \circ f)^{-1}(E)$ if and only if $g(f(x)) \in E$ if and only if $f(x) \in g^{-1}(E)$ which is obviously equivalent to $x \in f^{-1}(g^{-1}(E))$. Hence the identity holds and we complete the proof of the problem. ∎

Problem 1.7

$(\star)$ *Let p be a prime. For every $x, y \in \mathbb{Z}$, define the relation $x \equiv_p y$ to be such that $x - y$ is divisible by p. Prove that $\equiv_p$ is an equivalence relation.*

Proof. For every $x \in \mathbb{Z}$, since $x - x = 0$ is divisible by p, we have $x \equiv_p x$. If $x - y$ is divisible by p, then so is $y - x$ because $y - x = -(x - y)$. Thus we have $y \equiv_p x$ if $x \equiv_p y$. Finally, if $x - y = pM$ and $z - y = pN$ for some integers M and N, then we have

$$z - x = (z - y) + (y - x) = pN + pM = p(M + N).$$

In other words, we have $z \equiv_p x$ whenever $x \equiv_p y$ and $y \equiv_p z$. In conclusion, $\equiv_p$ is an equivalence relation and this completes the proof of the problem. ∎

1.3 Mathematical Induction

Problem 1.8

$(\star)$ *Prove that*

$$\frac{1}{2} \times \frac{3}{4} \times \cdots \times \frac{2n-1}{2n} < \frac{1}{\sqrt{2n+1}}$$

for every positive integer n.

Proof. When $n = 1$, since $\frac{1}{2} < \frac{1}{\sqrt{3}}$, the inequality holds in this case. Assume that

$$\frac{1}{2} \times \frac{3}{4} \times \cdots \times \frac{2k-1}{2k} < \frac{1}{\sqrt{2k+1}}$$

for some positive integer k. When $n = k + 1$, we have

$$\frac{1}{2} \times \frac{3}{4} \times \cdots \times \frac{2k-1}{2k} \times \frac{2k+1}{2k+2} < \frac{1}{\sqrt{2k+1}} \times \frac{2k+1}{2k+2} = \frac{\sqrt{2k+1}}{2k+2}. \tag{1.11}$$

Since $4k^2 + 8k + 3 < 4k^2 + 8k + 4$, we have $(2k+1)(2k+3) < (2k+2)^2$ which implies that

$$\frac{\sqrt{2k+1}}{2k+2} < \frac{1}{\sqrt{2k+3}}. \tag{1.12}$$

Combining the inequalities (1.11) and (1.12), we obtain

$$\frac{1}{2} \times \frac{3}{4} \times \cdots \times \frac{2k-1}{2k} \times \frac{2k+1}{2k+2} < \frac{1}{\sqrt{2k+3}}.$$

Hence the inequality is true for $n = k+1$ when it is true for $n = k$. By induction, we know that the inequality holds for all positive integers n. We end the proof of the problem. ∎

Problem 1.9

(⋆) *Use induction to prove Bernoulli's inequality*

$$(1 + x_1)(1 + x_2) \cdots (1 + x_n) \geq 1 + x_1 + \cdots + x_n,$$

where $x_1, x_2, \ldots, x_n$ have the same sign and $x_1, x_2, \ldots, x_n \geq -1$.

Proof. It is obvious that the statement is true for $n = 1$. Assume that

$$(1 + x_1)(1 + x_2) \cdots (1 + x_k) \geq 1 + x_1 + \cdots + x_k \tag{1.13}$$

for some positive integer k. When $n = k+1$, since $1 + x_{k+1} > 0$, we deduce from the assumption (1.13) that

$$(1 + x_1)(1 + x_2) \cdots (1 + x_k)(1 + x_{k+1})$$
$$\geq (1 + x_1 + \cdots + x_k)(1 + x_{k+1})$$
$$= 1 + x_1 + \cdots + x_k + x_{k+1} + (x_1 x_{k+1} + \cdots + x_k x_{k+1}). \tag{1.14}$$

Since $x_1, x_2, \ldots, x_{k+1}$ have the same sign, we have $x_i x_{k+1} \geq 0$ for all $i = 1, \ldots, k$. Therefore, we follow from this that $x_1 x_{k+1} + \cdots + x_k x_{k+1} \geq 0$ and then we yield from the inequality (1.14) that

$$(1 + x_1)(1 + x_2) \cdots (1 + x_k)(1 + x_{k+1}) \geq 1 + x_1 + \cdots + x_k + x_{k+1}.$$

Hence the statement is true for $n = k+1$ when it is true for $n = k$. By induction, we know that the inequality holds for all positive integers n. We complete the proof of the problem. ∎

Problem 1.10

$\star$ $\star$ *The Fibonacci numbers $\{F_n\}$ are defined by $F_1 = F_2 = 1$ and $F_{n+2} = F_{n+1} + F_n$, where $n \in \mathbb{N}$. Verify that*

$$F_n = \frac{1}{2^n\sqrt{5}}[(1+\sqrt{5})^n - (1-\sqrt{5})^n] \tag{1.15}$$

for every $n \in \mathbb{N}$.

Proof. It is easy to show that the formula (1.15) holds for $n = 1$ and $n = 2$. Assume that

$$F_k = \frac{1}{2^k\sqrt{5}}[(1+\sqrt{5})^k - (1-\sqrt{5})^k] \quad \text{and} \quad F_{k+1} = \frac{1}{2^{k+1}\sqrt{5}}[(1+\sqrt{5})^{k+1} - (1-\sqrt{5})^{k+1}]$$

for some positive integer k. For $n = k+2$, we obtain from the definition that

$$\begin{aligned}
F_{k+2} &= F_{k+1} + F_k \\
&= \frac{1}{2^k\sqrt{5}}[(1+\sqrt{5})^k - (1-\sqrt{5})^k] + \frac{1}{2^{k+1}\sqrt{5}}[(1+\sqrt{5})^{k+1} - (1-\sqrt{5})^{k+1}] \\
&= \frac{1}{2^{k+1}\sqrt{5}}[2(1+\sqrt{5})^k - 2(1-\sqrt{5})^k + (1+\sqrt{5})^{k+1} - (1-\sqrt{5})^{k+1}] \\
&= \frac{1}{2^{k+1}\sqrt{5}}[(1+\sqrt{5})^k(3+\sqrt{5}) - (1-\sqrt{5})^k(3-\sqrt{5})] \\
&= \frac{1}{2^{k+1}\sqrt{5}}\Big[(1+\sqrt{5})^k \times \frac{(1+\sqrt{5})^2}{2} - (1-\sqrt{5})^k \times \frac{(1-\sqrt{5})^2}{2}\Big] \\
&= \frac{1}{2^{k+2}\sqrt{5}}[(1+\sqrt{5})^{k+2} - (1-\sqrt{5})^{k+2}].
\end{aligned}$$

Hence the statement is true for $n = k+2$ when it is true for $n = k+1$ and $n = k$. By induction, we know that the formula (1.15) holds for all positive integers n. We complete the proof of the problem. ∎

CHAPTER 2

The Real Number System

▌2.1 Fundamental Concepts

The set of real numbers include all the rational numbers (e.g. the integer -4 and the fraction $\frac{1}{2}$) and all the irrational numbers (e.g. $\sqrt{3}$). Besides, a real number can also be thought of as a point lying on an infinitely long line called the **real number line**. In this section, fundamental properties regarding real numbers are reviewed and the main references here are [5, §2.2 - §2.4], [6, Chap. 1] [13, §1.12 - §1.22] and [15, Chap. 4 and 8].

▌2.1.1 Rational Numbers, Irrational Numbers and Real Numbers

The set of all rational numbers $\mathbb{Q}$ is defined by

$$\mathbb{Q} = \left\{ x \in \mathbb{R} \,\middle|\, x = \tfrac{p}{q}, \text{ where } p, q \in \mathbb{Z} \text{ and } q \neq 0 \right\}.$$

Recall that the set of all irrational numbers is denoted by $\mathbb{R} \setminus \mathbb{Q}$.

▌2.1.2 Absolute Values

Given $a \in \mathbb{R}$, the **absolute value** of a is defined by

$$|a| = \begin{cases} a, & \text{if } a \geq 0; \\ -a, & \text{if } a < 0. \end{cases}$$

It has the following useful properties:

- For every $a \in \mathbb{R}$, we have $|a| \geq 0$.

- $|a| = 0$ if and only if $a = 0$.

- For every $a, b \in \mathbb{R}$, we have $|a \times b| = |a| \times |b|$.

- $|a + b| \leq |a| + |b|$.

- $|a - b|$ can be treated as the **distance** between the two real numbers a and b.

2.1.3 The Completeness Axiom

What axioms "characterize" the set $\mathbb{R}$? In fact, there are three axioms which characterize $\mathbb{R}$. They are the "field axioms", the "order axioms" and the "completeness axiom". For details of the first two axioms, please refer to [13, §1.12 - §1.18, pp. 5 - 8]. The last axiom (completeness axiom) is a bit nontrivial:

The Completeness Axiom. *[3, §1.10] Every nonempty subset of $\mathbb{R}$ that is bounded above has a least upper bound (or a supremum) in $\mathbb{R}$.*

There are two useful results about $\mathbb{R}$ which can be deduced from the Completeness Axiom. The first one is known as the **Archimedean Property:**[a]

Theorem 2.1 (The Archimedean Property). *If $x, y \in \mathbb{R}$ and $x > 0$, then there exists a $n \in \mathbb{N}$ such that $nx > y$.*

Another important result is about the *density* of $\mathbb{Q}$ in $\mathbb{R}$.

Theorem 2.2 (Density of Rationals). *If $x, y \in \mathbb{R}$ and $x < y$, then there exists a $r \in \mathbb{Q}$ such that $x < r < y$, i.e.,*

$$\mathbb{Q} \cap (x, y) \neq \varnothing.$$

2.2 Rational and Irrational Numbers

Problem 2.1

⊛ *Show that $\sqrt{7}$ is irrational.*

Proof. Assume that $\sqrt{7}$ was rational. We write $\sqrt{7} = \frac{m}{n}$, where m and n are integers, $n \neq 0$ and they have no common divisor other than 1. Then we have

$$7n^2 = m^2. \tag{2.1}$$

- **Case (1): n is even.** Then $7n^2$ is also even. By the equation (2.1) and this fact, we have m^2 and thus m is even. However, this contradicts to the fact that m and n have no common divisor other than 1.

- **Case (2): n is odd.** In this case, $7n^2$ is odd and the equation (2.1) shows that m^2 and then m is also odd. Let $m = 2p + 1$ and $n = 2q + 1$ for some integers p and q. Now the equation (2.1) becomes

$$28q^2 + 28q + 7 = 4p^2 + 4p + 1. \tag{2.2}$$

  After simplification, the equation (2.2) reduces to

$$14q^2 + 14q + 3 = 2p^2 + 2p. \tag{2.3}$$

  We note that $14p^2 + 14p + 3$ is always odd, but $2p^2 + 2p$ is even. Therefore, the equation (2.3) cannot hold.

---

[a]For proofs of Theorems 2.1 and 2.2, please read [13, Theorem 1.20].

Hence we conclude from these that $\sqrt{7}$ is irrational, completing the proof of the problem. ∎

Problem 2.2

(⋆) Prove that $\sqrt{7} + \sqrt{2}$ is irrational.

Proof. Assume that $\sqrt{7} + \sqrt{2}$ was rational. By the fact that

$$(\sqrt{7} + \sqrt{2})(\sqrt{7} - \sqrt{2}) = 5,$$

the number $\sqrt{7} - \sqrt{2}$ is also rational. Thus the number $\sqrt{7}$ is also rational which contradicts Problem 2.1. Hence $\sqrt{7} + \sqrt{2}$ is irrational and this completes the proof of the problem. ∎

Problem 2.3

(⋆)(⋆) Suppose that n is a positive integer and $\theta = \sqrt{n} \in \mathbb{R}^+ \setminus \mathbb{Q}$. Prove that there exists a positive constant A such that

$$|p\theta - q| > \frac{A}{p} \tag{2.4}$$

for all integers p and q such that $p > 0$.

Proof. Let p and q be integers such that $p > 0$. Note that

$$(p\theta - q)(-p\theta - q) = -p^2\theta^2 + q^2 = -p^2 n + q^2$$

is an integer. Furthermore, $-p^2 n + q^2 \neq 0$; otherwise, $\theta = \sqrt{n} = \frac{q}{p}$, a contradiction. Therefore, it must be the case that

$$|p\theta - q| \times |p\theta + q| \geq 1 \tag{2.5}$$

Let $A = \min(\theta, \frac{1}{4\theta})$. This number is well-defined because $\theta \neq 0$. If $|p\theta - q| < p\theta$, then it follows from the triangle inequality that

$$|p\theta + q| = |-2p\theta + (p\theta - q)| \leq |2p\theta| + |p\theta - q| < 3p\theta. \tag{2.6}$$

Therefore, we know from the inequalities (2.5) and (2.6) that

$$|p\theta - q| \geq \frac{1}{|p\theta + q|} > \frac{1}{3p\theta} > \frac{A}{p}.$$

If $|p\theta - q| \geq p\theta$, then recall that $p \in \mathbb{N}$ so that

$$|p\theta - q| \geq p\theta \geq Ap > \frac{A}{p}.$$

Hence we have established the required inequality (2.4) and this completes the proof of the problem. ∎

Problem 2.4 (Dirichlet's Approximation Theorem)

$(\star)(\star)$ *Suppose that θ is irrational and N is a positive integer. Prove that there exist integers h, k with $0 < k \leq N$ such that*

$$|k\theta - h| < \frac{1}{N}. \tag{2.7}$$

Proof. Since θ is irrational, we have $i\theta \notin \mathbb{Z}$ for every nonzero integer i. By this fact, for $i = 0, 1, \ldots, N$, the $N + 1$ *distinct* numbers

$$t_i = i\theta - [i\theta]$$

belong to $[0, 1)$, where $[x]$ is the greatest integer less than or equal to x. By the Pigeonhole Principle,[b] there exist *at least* one interval $[\frac{j}{N}, \frac{j+1}{N})$ containing the numbers t_r and t_s, where $j = 0, 1, \ldots, N - 1$ and $r, s \in \{0, 1, \ldots, N\}$. Without loss of generality, we may assume that $r > s$ and

$$\frac{j}{N} \leq t_r - t_s < \frac{j+1}{N} \tag{2.8}$$

for some $j = 0, 1, \ldots, N - 1$. By the definition of t_i, we have

$$t_r - t_s = (r - s)\theta - ([r\theta] - [s\theta]) = k\theta - h, \tag{2.9}$$

where $h = [r\theta] - [s\theta]$ and $k = r - s$. It is clear that both h and k are integers with $0 < k \leq N$. By these and putting the identity (2.9) into the inequality (2.8), we obtain the result (2.7). This completes the proof of the problem. ∎

2.3   Absolute Values

Problem 2.5

$(\star)$ *For every $x, y \in \mathbb{R}$, prove that $||x| - |y|| \leq |x - y|$.*

Proof. By the triangle inequality, we have $|x| = |x - y + y| \leq |x - y| + |y|$ so that

$$|x| - |y| \leq |x - y| \tag{2.10}$$

Exchanging the roles of x and y in the inequality (2.10), we have $|y| - |x| \leq |x - y|$ and this implies that

$$|x| - |y| \geq -|x - y|. \tag{2.11}$$

Combining the inequalities (2.10) and (2.11), we obtain the desired inequality. This ends the proof of the problem. ∎

Problem 2.6

$(\star)$ *Solve the inequality $|x - 3| + |x + 3| \leq 14$.*

---

[b]If n items are put into m containers, where $n > m$, then *at least* one container contains more than one item.

Proof. Taking square to both sides of the inequality, we have

$$x^2 - 6x + 9 + x^2 + 6x + 9 + 2|x^2 - 9| \leq 196$$
$$x^2 + 9 + |x^2 - 9| \leq 98. \tag{2.12}$$

If $x^2 \leq 9$, then the inequality (2.12) becomes $18 \leq 98$ which is always true. Thus it means that the inequality holds when

$$|x| \leq 3. \tag{2.13}$$

If $x^2 > 9$, then we deduce from the inequality (2.12) that $x^2 \leq 49$. In this case, we have

$$3 < |x| \leq 7. \tag{2.14}$$

By the results (2.13) and (2.14), we know that the inequality holds for $-7 \leq x \leq 7$. We complete the proof of the problem. ∎

Problem 2.7

$(\star)$ *Suppose that a and b are real numbers. Prove that*

$$\frac{|a + b|}{1 + |a + b|} \leq \frac{|a|}{1 + |a|} + \frac{|b|}{1 + |b|}.$$

Proof. Since $0 \leq |a + b| \leq |a| + |b|$, we have $1 \leq 1 + |a + b| \leq 1 + |a| + |b|$ and then

$$\begin{aligned}
\frac{|a + b|}{1 + |a + b|} &= \frac{1 + |a + b| - 1}{1 + |a + b|} \\
&= 1 - \frac{1}{1 + |a + b|} \\
&\leq 1 - \frac{1}{1 + |a| + |b|} \\
&= \frac{|a| + |b|}{1 + |a| + |b|} \\
&\leq \frac{|a|}{1 + |a|} + \frac{|b|}{1 + |b|},
\end{aligned}$$

completing the proof of the problem. ∎

2.4 The Completeness Axiom

Problem 2.8

$(\star)(\star)$ *Prove that $\mathbb{N}$ has no upper bound.*

Proof. Assume that $\mathbb{N}$ had an upper bound. By the Completeness Axiom, we know that $\sup \mathbb{N}$ exists in $\mathbb{R}$. By definition, $\sup \mathbb{N} - 1$ is not an upper bound of $\mathbb{N}$, so there exists $n \in \mathbb{N}$ such that

$$\sup \mathbb{N} - 1 < n. \tag{2.15}$$

Since $n + 1 \in \mathbb{N}$, we also have

$$n + 1 \leq \sup \mathbb{N}. \tag{2.16}$$

It follows from the inequalities (2.15) and (2.16) that

$$\sup \mathbb{N} < n + 1 \leq \sup \mathbb{N}$$

which is a contradiction. Hence $\mathbb{N}$ has no upper bound and this ends the proof of the problem. ∎

Problem 2.9

(⋆)(⋆) *Let $a, b \in \mathbb{R}$ and $a < b$. Prove that there is an irrational number θ such that $a < \theta < b$.*

Proof. Since $a < b$, Theorem 2.2 (Density of Rationals) implies that there exists a $r \in \mathbb{Q}$ such that $a < r < b$. Since $b - r > 0$ and $\sqrt{2} > 0$, Theorem 2.1 (The Archimedean Property) ensures that one can find a positive integer n such that $n(b - r) > \sqrt{2} > 0$. Note that $\frac{\sqrt{2}}{n}$ is irrational and so the number

$$\theta = r + \frac{\sqrt{2}}{n}$$

is also irrational. It is easy to see that

$$a < r < \theta = r + \frac{\sqrt{2}}{n} < b.$$

Hence we complete the proof of the problem. ∎

Remark 2.1

Problem 2.9 can be rephrased as the density of irrational numbers. There is another proof which uses the fact that the interval (a, b) is uncountable, see Problem 3.1.

Problem 2.10

(⋆)(⋆) *Let $x \geq 0$ and n be a positive integer. Show that there exists a unique $y \geq 0$ such that*

$$y^n = x. \tag{2.17}$$

Proof. If $x = 0$, then we take $y = 0$. Thus we may suppose that $x > 0$. Furthermore, the case $n = 1$ is trivial, so we may assume that $n \geq 2$.

We first prove the uniqueness part of this problem. If $y_1 > 0$ and $y_2 > 0$ satisfy the equation (2.17), then we have

$$0 = y_1^n - y_2^n = (y_1 - y_2)(y_1^{n-1} + y_1^{n-2}y_2 + \cdots + y_1 y_2^{n-2} + y_2^{n-1}), \tag{2.18}$$

where n is a positive integer. Since y_1 and y_2 are positive, $y_1^{n-1} + y_1^{n-2}y_2 + \cdots + y_1 y_2^{n-2} + y_2^{n-1}$ must be positive. Therefore, we deduce from the equation (2.18) that $y_1 = y_2$. This proves the uniqueness.

Now we show the existence of the y as follows: Suppose that

$$E = \{y \geq 0 \,|\, y^n \leq x\}$$

which is a subset of $\mathbb{R}$. Since $x > 0$, Theorem 2.1 (The Archimedean Property) guarantees the existence of a positive integer m such that $mx > 1$. Since $m \in \mathbb{N}$, we have

$$0 < \left(\frac{1}{m}\right)^n < \frac{1}{m} < x$$

for every $n \geq 2$. This shows that $\frac{1}{m} \in E$. Next, if p is a positive integer such that $p \geq x$, then we have $y^n \leq x \leq p$ for every $y \in E$. Thus we have shown that E is a nonempty subset of $\mathbb{R}$ that is bounded above. Hence the Completeness Axiom implies that $\alpha = \sup E$ exists in $\mathbb{R}$. This number must be positive because E contains the positive number $\frac{1}{m}$.

Assume that $\alpha^n < x$. Now for each $p \in \mathbb{N}$, we have

$$\left(\alpha + \frac{1}{p}\right)^n = \alpha^n + \sum_{k=1}^{n} C_k^n \alpha^{n-k} p^{-k} \leq \alpha^n + p^{-1} \sum_{k=1}^{n} C_k^n \alpha^{n-k}. \tag{2.19}$$

Denote $A = \sum_{k=1}^{n} C_k^n \alpha^{n-k}$. Then it follows from Theorem 2.1 (The Archimedean Property) that there is a positive integer q such that $q(x - \alpha^n) > A$, i.e.,

$$x - \alpha^n > \frac{A}{q}.$$

We substitute $p = q$ in the inequality (2.19) to obtain

$$\left(\alpha + \frac{1}{q}\right)^n \leq \alpha^n + \frac{A}{q} < \alpha^n + x - \alpha^n = x.$$

By definition, $\alpha + \frac{1}{q} \in E$ so that $\alpha + \frac{1}{q} \leq \alpha$ which is a contradiction. Thus the case $\alpha^n < x$ is impossible. The impossibility of the other case $\alpha^n > x$ can be shown similarly. Hence we conclude that $\alpha^n = x$ and this completes the proof of the problem. ∎

Problem 2.11

(⋆) Let E be a nonempty subset of $\mathbb{R}$ having least upper bounds α and β. Prove that $\alpha = \beta$.

Proof. Since α and β are least upper bounds of E, they are upper bounds of E. By definition, both inequalities

$$\alpha \leq \beta \quad \text{and} \quad \beta \leq \alpha$$

hold simultaneously. Hence we definitely have $\alpha = \beta$, completing the proof of our problem. ∎

Problem 2.12

$\star$ $\star$ Let E be a nonempty subset of $\mathbb{R}$ and $y = \sup E$. Let

$$-E = \{-x \in \mathbb{R} \,|\, x \in E\}.$$

Prove that $-y$ is the greatest lower bound of $-E$.

Proof. Since $x \leq y$ for all $x \in E$, we have $-y \leq -x$ for all $x \in E$. Therefore, $-y$ is a lower bound of $-E$. Assume that there was $w \in \mathbb{R}$ such that

$$-y < w \leq z \tag{2.20}$$

for all $z \in -E$. Since $z = -x$ for some $x \in E$, we deduce from the inequalities (2.20) that $-y < w \leq -x$ or equivalently,

$$y > -w \geq x \tag{2.21}$$

for all $x \in E$. However, the inequalities (2.21) say that $y \neq \sup E$, a contradiction. Hence we must have $-y = \inf(-E)$, completing the proof of the problem. ∎

Problem 2.13

$\star$ $\star$ Suppose that E and F are two nonempty subsets of positive real numbers and they are bounded above. Let

$$EF = \{x \times y \,|\, x \in E \text{ and } y \in F\}.$$

Prove that $\sup(EF) = \sup E \times \sup F$.

Proof. Since E and F are nonempty and bounded above, EF is also nonempty and bounded above. By the Completeness Axiom, $\sup E$, $\sup F$ and $\sup(EF)$ exist in $\mathbb{R}^+$. It is clear that $\sup E \times \sup F$ is an upper bound of EF, i.e.,

$$\sup(EF) \leq \sup E \times \sup F. \tag{2.22}$$

Now we are going to show that the equality in (2.22) holds. Given that $\epsilon > 0$. We claim that there exist $x_0 \in E$ and $y_0 \in F$ such that

$$\sup E \times \sup F - \epsilon < x_0 y_0.$$

To this end, since $\sup E >$ and $\sup F > 0$, we can find a large positive integer N such that

$$\sup E - \frac{\epsilon}{N} > 0, \quad \sup F - \frac{\epsilon}{N} > 0 \quad \text{and} \quad \sup E + \sup F < N. \tag{2.23}$$

By the definition of supremum, we must have $\sup E - \frac{\epsilon}{N} < x_0$ and $\sup F - \frac{\epsilon}{N} < y_0$ *for some* $x_0 \in E$ and $y_0 \in F$, so they imply that

$$\sup E \times \sup F - \frac{\epsilon}{N}\left(\sup E + \sup F - \frac{\epsilon}{N}\right) = \left(\sup E - \frac{\epsilon}{N}\right)\left(\sup F - \frac{\epsilon}{N}\right) < x_0 y_0. \tag{2.24}$$

Now we are able to derive from the inequalities (2.23) and (2.24) that

$$\sup E \times \sup F - \epsilon < \sup E \times \sup F - \frac{\epsilon}{N}\left(\sup E + \sup F\right)$$
$$< \sup E \times \sup F - \frac{\epsilon}{N}\left(\sup E + \sup F - \frac{\epsilon}{N}\right)$$
$$< x_0 y_0$$

for some $x_0 y_0 \in EF$. Hence this proves our claim and it means that $\sup E \times \sup F$ is in fact the least upper bound of EF, i.e., $\sup(EF) = \sup E \times \sup F$ and this completes the proof of the problem. ∎

Problem 2.14

(⋆) *Let p and q be two primes. Let $E = \{p^{-s} + q^{-t} \,|\, s, t \in \mathbb{N}\}$. What are $\sup E$ and $\inf E$?*

Proof. Since $0 < \frac{1}{p^s} \leq \frac{1}{p}$ and $0 < \frac{1}{q^t} \leq \frac{1}{q}$ for every $s, t \in \mathbb{N}$, we have

$$0 < x \leq \frac{1}{p} + \frac{1}{q}$$

for every $x \in E$. Since $\frac{1}{p} + \frac{1}{q} \in E$, it is the maximum of E. In other words, we have

$$\sup E = \max E = \frac{1}{p} + \frac{1}{q}.$$

Let $\epsilon > 0$. We know from Theorem 2.2 that there exists a $s_0 \in \mathbb{N}$ such that $0 < \frac{1}{p^{s_0}} < \frac{\epsilon}{2}$. Similarly, by Theorem 2.2 again, there exists a $t_0 \in \mathbb{N}$ such that $0 < \frac{1}{q^{t_0}} < \frac{\epsilon}{2}$. Therefore, it is clear that they give

$$0 < \frac{1}{p^{s_0}} + \frac{1}{q^{t_0}} < \epsilon$$

which implies that ϵ is not a lower upper of E. Hence we have $\inf E = 0$ and we finish the proof of the problem. ∎

Problem 2.15

(⋆)(⋆) *Let*

$$E = \left\{\frac{p}{q} \,\middle|\, p, q \in \mathbb{N} \text{ and } 0 < p < q\right\}.$$

Prove that E does not contain a maximum element or a minimum element. Find $\sup E$ and $\inf E$.

Proof. Since $q > p > 0$, we have $pq + q > pq + p > 0$ and then

$$\frac{p+1}{q+1} > \frac{p}{q}.$$

Thus E does not contain the maximum element. Similarly, since $0 < \frac{p}{q} < 1$, we must have

$$0 < \frac{p^2}{q^2} < \frac{p}{q}$$

which implies that E does not contain the minimum element.

To find $\sup E$ and $\inf E$, we first note that

$$0 < \frac{p}{q} < 1$$

for every $\frac{p}{q} \in E$. Therefore, 1 and 0 are an upper bound and a lower bound of E. If $\sup E \neq 1$, then we can find a rational $\frac{m}{n}$ such that

$$\frac{p}{q} < \frac{m}{n} < 1 \tag{2.25}$$

for every $\frac{p}{q} \in E$. Here m and n must satisfy $0 < m < n$. In other words, we have $\frac{m}{n} \in E$. By this and the inequality (2.25), E contains the maximum element $\frac{m}{n}$, a contradiction. Hence we establish the result that

$$\sup E = 1.$$

The case for $\inf E = 0$ is similar and so we omit the details here. We have completed the proof of the problem. ∎

CHAPTER 3

Countable and Uncountable Sets

▌ 3.1 Fundamental Concepts

When we are given a set containing a *finite* number of objects, then we can simply count the number of objects and determine the **cardinality** of that set. Can we do a similar thing when the set has *infinitely* many objects? The answer to this question is "yes" if we define the concept of **countability** of such an infinite set. In fact, countability of a set A is to "measure how many" elements the set A contains. The main references for this chapter are [3, §2.12 & §2.15], [13, §2.8, §2.12 & §2.13] and [16, §2.4].

▌ 3.1.1 Definitions of Countable and Uncountable Sets

Definition 3.1. *Given two sets A and B. If there exists a function $f : A \to B$ which is one-to-one and onto, then we say that A and B **equivalent**.*

It can be shown easily that this is indeed an equivalent relation, see §1.1.3. Now we are ready to apply this equivalent relation to define the countability of a set E. Since there are two sets in Definition 3.1, we need to choose a set to compare with A. The most natural one is the set of all positive integers $\mathbb{N}$.

Definition 3.2 (Countability). *A set A is said to be **countable** if it is equivalent to $\mathbb{N}$. In other words, there exists a function $f : \mathbb{N} \to A$ which is one-to-one and onto. If A is not finite or countable, then it is **uncountable**.*

Definition 3.3. *We say that A is **at most countable** if A is either finite or countable.*

▌ 3.1.2 Properties of Countable Sets

Here we list some important and useful results about the countability of a set A.

Theorem 3.4. *Every infinite subset of a countable set A must be countable*

19

Theorem 3.5. *Suppose that* $\{A_1, A_2, \ldots\}$ *is a countable collection of countable sets. Then the set*

$$A = \bigcup_{k=1}^{\infty} A_k \tag{3.1}$$

is countable.

Theorem 3.6. *Suppose that* $\{A_1, A_2, \ldots, A_n\}$ *is a finite collection of countable sets. Then the set*

$$A = A_1 \times A_2 \times \cdots \times A_n$$

is countable.

As an immediate application of Theorem 3.6, we see that the sets $\mathbb{Z}$ and $\mathbb{Q}$ are countable. However, we note that Theorem 3.6 cannot hold for *infinite* collection of countable (or even finite) sets, see Problem 3.10 for a counterexample.

3.2 Problems on Countable and Uncountable Sets

Problem 3.1

⋆ *Reprove Problem 2.9 by using the fact that* (a, b) *is uncountable.*

Proof. Since $\mathbb{Q} \cap (a, b) \subset \mathbb{Q}$, Theorem 3.4 implies that it is countable. Since (a, b) is given uncountable, the set

$$(a, b) \setminus [\mathbb{Q} \cap (a, b)] \tag{3.2}$$

must be uncountable. Since elements in the set (3.2) are *all* irrational numbers in (a, b), there exists an irrational number θ such that $a < \theta < b$. This completes the proof of the problem. ∎

Problem 3.2

⋆ *Suppose that* $(0, 1)$ *is uncountable. Is* $\mathbb{R}$ *uncountable?*

Proof. The answer is affirmative. Assume that $\mathbb{R}$ was countable. Since $(0, 1) \subset \mathbb{R}$, $(0, 1)$ is also countable by Theorem 3.4, a contradiction to the hypothesis. Hence we have $\mathbb{R}$ is uncountable. We end the proof of the problem. ∎

Problem 3.3 (Cantor's Theorem)

⋆⋆ *Suppose that* A *is a nonempty set. Denote* 2^A *to be the **power set** of* A *(the set of all subsets of* A*). Show that there does not exist a surjective function* $f : A \to 2^A$.

Proof. Assume that there was a surjective function $f : A \to 2^A$. We consider the subset $B \subseteq A$ defined by

$$B = \{x \in A \mid x \notin f(x)\}. \tag{3.3}$$

It is clear that $B \subset A$ so that $B \in 2^A$. Since f is surjective, there exists a unique $x_0 \in A$ such that

$$f(x_0) = B.$$

Does $x_0 \in B$? On the one hand, if $x_0 \in B$, then the definition (3.3) gives $x_0 \notin f(x_0) = B$, a contradiction. On the other hand, if $x_0 \notin B$, then the definition (3.3) again implies that $x_0 \in f(x_0) = B$ which is a contradiction again. Hence these contradictions show that the existence of such element x_0 and then the existence of such surjective function f is impossible, finishing the proof of the problem. ∎

Problem 3.4

(⋆) Prove that $2^{\mathbb{N}}$ is uncountable.

Proof. Assume that $2^{\mathbb{N}}$ was countable. Then there exists an one-to-one and onto function

$$f : \mathbb{N} \to 2^{\mathbb{N}}$$

which certainly contradicts the result of Problem 3.3. Hence $2^{\mathbb{N}}$ is uncountable and this completes the proof of the problem. ∎

Problem 3.5

(⋆)(⋆) Prove Theorem 3.5.

Proof. Let A be defined as in the equation (3.1). For each k, since A_k is countable, we can list its elements as

$$A_k = \{a_{k1}, a_{k2}, \ldots\}.$$

Consider the following array:

$$
\begin{array}{llll}
A_1: & a_{11} & a_{12} & a_{13} & \cdots \\
A_2: & a_{21} & a_{22} & a_{23} & \cdots \\
A_3: & a_{31} & a_{32} & a_{33} & \cdots \\
\cdots & \cdots\cdots\cdots\cdots\cdots &
\end{array}
$$

As indicated by the colored elements, we see that they can be arranged in a sequence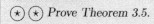

$$a_{11}, a_{21}, a_{12}, a_{31}, a_{22}, \ldots .a_{13}, \ldots \tag{3.4}$$

If x is a common element of two distinct sets A_i and A_j, then it appears more than once in the sequence (3.4). Therefore there exists a subset $B \subseteq \mathbb{N}$ and a function $f : B \to A$ such that f is one-to-one and onto. By Definition 3.3, A is at most countable. Since $A_1 \subseteq A$ and A_1 is countable, A is an infinite set and we see from this that A is also countable. This ends the proof of the problem. ∎

Problem 3.6

$(\star)$ *Prove that the set $E = \{p + q\sqrt{3} \mid p, q \in \mathbb{Q}\}$ is countable.*

Proof. Since $\mathbb{Q}$ is countable, $\mathbb{Q} \times \mathbb{Q}$ is also countable by Theorem 3.6. Define $f : \mathbb{Q} \times \mathbb{Q} \to E$ by

$$f(p,q) = p + q\sqrt{3}.$$

We see easily that f is surjective, i.e., $E = f(\mathbb{Q} \times \mathbb{Q})$. Note that this function is also injective because if $(p_1, q_1), (p_2, q_2)$ are two distinct elements in E such that $f(p_1, q_1) = f(p_2, q_2)$. Then we have

$$p_1 + q_1\sqrt{3} = p_2 + q_2\sqrt{3}$$
$$\sqrt{3} = \frac{p_1 - p_2}{q_2 - q_1}$$

which implies that $\sqrt{3}$ is irrational, a contradiction. Hence E is also countable by Definition 3.2 and this completes the proof of the problem. ∎

Problem 3.7

$(\star)$ *Prove that the set of straight lines passing through the origin in $\mathbb{R}^2$ is uncountable.*

Proof. Any line L passing through the origin in $\mathbb{R}^2$ must be in the form

$$y = mx \quad \text{or} \quad x = 0,$$

where $m \in \mathbb{R}$. Let this set be A. Since A has a subset B which is equivalent to $\mathbb{R}$, it follows from Problem 3.2 that B is uncountable. Assume that A was countable. Then every infinite subset of A must be countable by Theorem 3.4. In particular, B is countable which is a contradiction. Hence A must be uncountable and we finish the proof of the problem. ∎

Problem 3.8

$(\star)$ *Denote A to the set of circles in $\mathbb{R}^2$ having rational radii and centres with rational coordinates. Verify that A is countable.*

Proof. If a circle has rational radius r_1 and rational centre (r_2, r_3), where $r_1, r_2, r_3 \in \mathbb{Q}$, then it is *uniquely determined* by the triple (r_1, r_2, r_3). Therefore, A is a subset of $\mathbb{Q} \times \mathbb{Q} \times \mathbb{Q}$. By Theorem 3.6, $\mathbb{Q} \times \mathbb{Q} \times \mathbb{Q}$ is countable. Since $(n, n, n) \in A$ for every $n \in \mathbb{N}$, A must be infinite and by Theorem 3.4, it is countable. This completes the proof of the problem. ∎

Problem 3.9

$(\star)(\star)$ *Suppose that $F \subset E$, where E is uncountable and F is countable. Show that $E \setminus F$ is uncountable. Find $|E \setminus F|$, where $|E|$ means the cardinality of the set E.*

Proof. Note that $E = (E \setminus F) \cup F$. Thus if both $E \setminus F$ and F are countable, then Theorem 3.4 implies that E must be countable which is impossible. Hence this proves our first assertion that $E \setminus F$ is uncountable.

For the second assertion, we claim that

$$|E \setminus F| = |E|. \tag{3.5}$$

Since F is a proper subset of E, we have $E \setminus F \neq \varnothing$ and then we can find a $x_1 \in E \setminus F$. Now we consider the set $E \setminus (F \cup \{x_1\})$. Since F is countable, $F \cup \{x_1\}$ is also countable by Theorem 3.5 and we follow from the first assertion that $E \setminus (F \cup \{x_1\})$ is still uncountable. In particular, we must have $E \setminus (F \cup \{x_1\}) \neq \varnothing$. Therefore, there exists a $x_2 \in E \setminus (F \cup \{x_1\})$. Now this process can be done inductively and we can find a subset $X = \{x_1, x_2, \ldots\} \subset E \setminus F$, where all $x_i \neq x_j$ if $i \neq j$ and the set $E \setminus (X \cup F)$ is uncountable.

It is clear that both X and $X \cup F$ are countable, so we can find a bijective function $f : X \cup F \to X$ by Definition 3.2. We define $g : E \to E \setminus F$ by

$$g(x) = \begin{cases} x, & \text{if } x \in E \setminus (X \cup F); \\ f(x), & \text{if } x \in X \cup F. \end{cases} \tag{3.6}$$

If we can show that g is bijective, then we obtain the result (3.5). To this end, suppose that $g(x) = g(y)$. We claim that it is impossible to have either

"$x \in E \setminus (X \cup F)$ and $y \in X \cup F$" or "$y \in E \setminus (X \cup F)$ and $x \in X \cup F$".

In fact, if $x \in E \setminus (X \cup F)$ and $y \in X \cup F$, then we have $g(x) = x \notin X \cup F$ but $g(y) = f(y) \in X$, so we cannot have

$$g(x) = g(y)$$

and therefore the situation "$x \in E \setminus (X \cup F)$ and $y \in X \cup F$" is impossible. Similarly, the situation "$y \in E \setminus (X \cup F)$ and $x \in X \cup F$" can also be shown impossible. Now there are two remaining cases which are

- **Case (1):** $x, y \in E \setminus (X \cup F)$. In this case, we have $g(x) = x$ and $g(y) = y$ so that $x = y$.

- **Case (2):** $x, y \in X \cup F$. In this case, we have $g(x) = f(x)$ and $g(y) = f(y)$ so that $f(x) = f(y)$ which implies that $x = y$.

In conclusion, we have shown that g is injective.

Next we check that g is surjective. Let $y \in E \setminus F$. If $y \in X$, then we take $x = f^{-1}(y) \in X \cup F$ so that the definition (3.6) implies

$$g(x) = f(x) = y.$$

If $y \notin X$, then $y \in E \setminus (X \cup F)$. Thus we take $x = y$ and the definition (3.6) implies

$$g(x) = x = y.$$

In conclusion, we have shown that g is surjective. Hence, this proves our desired result (3.5), completing the proof of the problem. ∎

Problem 3.10

$\star$ $\star$ *For each $k \in \mathbb{N}$, define $A_k = \{0, 1\}$. Prove that the set*

$$A = \prod_{k=1}^{\infty} A_k = \{0, 1\} \times \{0, 1\} \times \cdots$$

is uncountable.

Proof. Assume that A was countable. By Definition 3.2, a bijective function $f : \mathbb{N} \to A$ exists. For every $n \in \mathbb{N}$, denote $f(n)_n \in \{0, 1\}$ to be the n-th "coordinate" of $f(n)$, i.e.,

$$f(n) = (*, *, \ldots, *, \underbrace{f(n)_n}_{n\text{-th coordinate}}, *, \ldots).$$

Next, we define $a = (a_1, a_2, \ldots, a_n, \ldots)$ by

$$a_n = 1 - f(n)_n$$

for each $n \in \mathbb{N}$. It is clear that we always have $a \in A$. However, $f(n) \neq a$ for every $n \in \mathbb{N}$ because the n-digits of $f(n)$ and a are different. This fact contradicts the surjective property of f. Hence A must be uncountable and this finishes the proof of the problem. $\blacksquare$

Problem 3.11

$\star$ $\star$ *A real number is said to be* **algebraic** *if it is a root of an equation*

$$a_n x^n + \cdots + a_1 x + a_0 = 0$$

with integer coefficients, where $a_n \neq 0$. Show that the set of all algebraic numbers, denoted by $\mathcal{A}$, is countable.

Proof. For each $n = 0, 1, 2, \ldots$, define

$$A_n = \{a_n x^n + \cdots + a_1 x + a_0 \,|\, a_n \in \mathbb{Z} \setminus \{0\}, a_0, a_1, \ldots, a_{n-1} \in \mathbb{Z}\}.$$

Next, we define $f : A_n \to (\mathbb{Z} \setminus \{0\}) \times \mathbb{Z}^n$ by

$$f(a_n x^n + \cdots + a_1 x + a_0) = (a_n, \ldots, a_1, a_0)$$

which is obviously bejective. Since $\mathbb{Z} \setminus \{0\}$ and $\mathbb{Z}$ are countable, Theorem 3.6 implies that $(\mathbb{Z} \setminus \{0\}) \times \mathbb{Z}^n$ is also countable. By Definition 3.2, each A_n is countable. By Theorem 3.5, the set

$$A = \bigcup_{n=0}^{\infty} A_n$$

is countable. Since $\mathcal{A}$ is equivalent to a subset of A and each integer is algebraic, $\mathcal{A}$ must be infinite and thus it is countable by Theorem 3.4. We complete the proof of the problem. $\blacksquare$

Problem 3.12

$(\star)$ *Let $\mathcal{A}$ be the set of all algebraic real numbers. Show that $\mathbb{R} \setminus \mathcal{A}$ is uncountable.*

Proof. By Problem 3.11, $\mathcal{A}$ is countable. If $\mathcal{A} = \mathbb{R}$, then $\mathbb{R}$ is also countable which contradicts Problem 3.2. Thus we must have $\mathcal{A} \subset \mathbb{R}$ and Problem 3.9 shows that $\mathbb{R} \setminus \mathcal{A}$ is uncountable. This completes the proof of the problem. ∎

Problem 3.13

$(\star)$ *Construct a bijection between the sets $(0, 1)$ and $(0, 1]$.*

Proof. Define $A = \{\frac{1}{2}, \frac{1}{3}, \ldots\} \subseteq (0, 1)$ and $B = \{1, \frac{1}{2}, \frac{1}{3}, \ldots\} \subseteq (0, 1]$. Define $f : A \to B$ by

$$f\left(\frac{1}{n}\right) = \frac{1}{n - 1}$$

which is clearly a bijection. Next, we define $F : (0, 1) \to (0, 1]$ by

$$F(x) = \begin{cases} f(x), & \text{if } x \in A; \\ x, & \text{if } x \in (0, 1) \setminus A. \end{cases}$$

Now it is easy to check that F is in fact a bijection, completing the proof of the problem. ∎

CHAPTER 4

Elementary Topology on Metric Spaces

4.1 Fundamental Concepts

In the real number line $\mathbb{R}$, everyone knows that we can find the **distance** between two points $a, b \in \mathbb{R}$ by simply evaluating the absolute value $|a - b|$ (see §2.1.2). This situation can be generalized without any difficulties. In fact, most of the properties of the real number $\mathbb{R}$ equipped with the absolute value $|\cdot|$ can be treated as special cases of the more general setting of a **metric space** with a **distance function** (or a **metric**). Therefore, instead of studying topological properties of the real number system $\mathbb{R}$, we choose to review some topological results of a metric space. For details of the theory, please read [3, Chap. 3] or [13, §2.15 - §2.47, pp. 30 - 43].

4.1.1 Metric Spaces

Suppose that X is a set. Now elements of X are called **points**.

Definition 4.1. *A metric space is a pair (X, d) of a set X and a function $d : X \times X \to \mathbb{R}$ satisfying the following conditions for any points $p, q, r \in X$:*

(1) $d(p, p) = 0$ and $d(p, q) > 0$ if $p \neq q$;

(2) $d(p, q) = d(q, p)$;

(3) $d(p, q) \leq d(p, r) + d(r, q)$.

Any function satisfying the conditions in Definition 4.1 is called a **distance function** or a **metric**. It is well-known that the set $\mathbb{R}^n$ is a metric space, where n is a positive integer. In the following discussion and problems, we assume that the notation X means a metric space with a metric d.

4.1.2 Open Sets and Closed Sets

Definition 4.2. *A **neighborhood** of a point $p \in X$ with **radius** $r > 0$, denoted by $N_r(p)$, is defined by*

$$N_r(p) = \{q \in X \mid d(p, q) < r\}.$$

Definition 4.3. *Let E be a set. A point $p \in E$ is called an* **interior point** *of E if $N_r(p) \subseteq E$ for some $r > 0$. We call E an* **open set** *in X if every point of E is an interior point of E.*

Theorem 4.4. *Let $p \in X$. Every neighborhood $N_r(p)$ is open in X.*

Definition 4.5. *Let E be a set. A point $p \in X$ is called a* **limit point** *of E if*

$$N_r(p) \cap (E \setminus \{p\}) \neq \varnothing$$

for every $r > 0$. Denote the set of all limit points of E to be E'. We call E a **closed set** *in X if $E' \subseteq E$.*

Theorem 4.6. *A set E is open in X if and only if E^c is closed in X.*

Theorem 4.7.

(a) *For any families $\{E_\alpha\}$ of open sets and $\{F_\alpha\}$ of closed sets, $\bigcup E_\alpha$ is open in X and $\bigcap F_\alpha$ is closed in X.*

(b) *For any finite families $\{E_1, E_2, \ldots, E_n\}$ of open sets and $\{F_1, F_2, \ldots, F_n\}$ of closed sets, $\bigcap\limits_{k=1}^{n} E_k$ is open in X and $\bigcup\limits_{k=1}^{n} F_k$ is closed in X.*

4.1.3 Interiors and Closures

Definition 4.8. *Let E be a subset of X. The set*

$$\overline{E} = E \cup E'$$

is called the **closure** *of E and the collection of all interior points of E is denoted by E° and it is called the* **interior** *of E.*

Theorem 4.9. *Given that $E \subseteq X$. We have*

(a) *$\overline{E}$ is closed in X.*

(b) *$E = \overline{E}$ if and only if E is closed.*

(c) *if $E \subseteq F$ and F is closed in X, then $\overline{E} \subseteq F$.*

By Theorem 4.9(a) and (c), we see that $\overline{E}$ is the **smallest** closed subset of X containing E. There is a similar result for the interior of E. In fact, we see from Definition 4.8 that an interior point of E lies in E so that we must have $E^\circ \subseteq E$. Then it can be shown that E° is the **largest** open subset of X inside E, see Problem 4.11.

Interiors and closures of sets have many properties in very "similar" forms. The following table summarizes and compares such properties of the interior and the closure of a set E in the metric space X.

	Interior of E	Closure of E
Subset relation	$E \subset X$ implies $E^\circ \subseteq X^\circ$	$E \subset X$ implies $\overline{E} \subseteq \overline{X}$
Finite union	$E^\circ \cup F^\circ \subseteq (E \cup F)^\circ$	$\overline{E \cup F} = \overline{E} \cup \overline{F}$
Finite intersection	$(E \cap F)^\circ = E^\circ \cap F^\circ$	$\overline{E \cap F} \subseteq \overline{E} \cap \overline{F}$
Arbitrary union	$\bigcup_\alpha E_\alpha^\circ \subseteq \left(\bigcup_\alpha E_\alpha \right)^\circ$	$\bigcup_\alpha \overline{E_\alpha} \subseteq \overline{\bigcup_\alpha E_\alpha}$
Arbitrary intersection	$\left(\bigcap_\alpha E_\alpha \right)^\circ \subseteq \bigcap_\alpha E_\alpha^\circ$	$\overline{\bigcap_\alpha E_\alpha} \subseteq \bigcap_\alpha \overline{E_\alpha}$
Finite product relation	$(E \times F)^\circ = E^\circ \times F^\circ$	$\overline{E \times F} = \overline{E} \times \overline{F}$

Table 4.1: Properties of the interior and the closure of E in X.

Proofs of the finite union and intersection of interiors can be seen in Problem 4.12 and a proof of the finite intersection of closures can be found in Problem 4.13.

4.1.4 Sets in Metric Subspaces

Let Y be a subset of a metric space X. Then it is natural to think that Y is also a metric space with the same metric as X because all points in Y satisfy Definition 4.1. Therefore, we have to pay particular attention when we talk about open (resp. closed) sets in X and open (resp. closed) sets in Y.

Let $E \subseteq Y \subseteq X$, where both X and Y are metric spaces. It may happen that E is open in Y, but *not* open in X. An example can be found in [13, Example 2.21(g), p. 33]. Fortunately, there is a test for E to be open or closed in X:

Theorem 4.10. *Let $E \subseteq Y \subseteq X$. Then E is open (resp. closed) in Y if and only if*

$$E = Y \cap F$$

for some open (resp. closed) subset F of X.

Is there any similar result for interiors and closures of subsets of X and of Y? The answer to this question is "yes" to closures, but "no" to interiors. In fact, we have the following result:

Theorem 4.11. *Given that $E \subseteq Y \subseteq X$. Denote E_X° and $\overline{E}_X$ to be the interior and the closure of E in X respectively. Similarly, E_Y° and $\overline{E}_Y$ are the interior and the closure of E in Y respectively. Then we have*

(a) $\overline{E}_Y = \overline{E}_X \cap Y$;

(b) $E_Y^\circ \supseteq E_X^\circ \cap Y$.

We remark that the inclusion in (b) can be proper, see Problem 4.15 for an example.

4.1.5 Compact Sets in Metric Spaces

Let K be a subset of X. A collection $\{V_\alpha\}$ of subsets of X is said to **cover** K if

$$K \subseteq \bigcup_\alpha V_\alpha.$$

By an **open cover** of K, we mean that each V_α is open in X.

Definition 4.12 (Compact Sets). *A subset K of X is said to be **compact** if for every open cover $\{V_\alpha\}$ of K, there are finitely many indices $\alpha_1, \ldots, \alpha_n$ such that*

$$K \subseteq V_{\alpha_1} \cup V_{\alpha_2} \cup \cdots \cup V_{\alpha_n}.$$

In particular, if the metric space X is itself compact, then we call it a **compact metric space**. There are many nice properties for compact sets. One of them is that it **does not** depend on the space in which K lies. More explicitly, this says that if $K \subseteq Y \subseteq X$, then K is compact in X if and only if K is compact in Y. See [13, Theorem 2.33, p. 37] for a proof of it. Other important and useful properties of compact sets are given as follows:

Theorem 4.13. *Let K be a compact subset of a metric space X.*

(a) *K is closed in X.*

(b) *If E is closed in K, then E is compact. In particular, $E \cap K$ is compact for every closed subset E of X.*

Another important property of compact sets is linked to the so-called **Cantor's intersection theorem** which concentrates on intersections of **decreasing nested sequences of non-empty compact sets**. In fact, the result is stated as follows:

Theorem 4.14. *Suppose that $\{K_\alpha\}$ is a family of compact subsets of X. If the intersection of every finite subcollection of $\{K_\alpha\}$ is nonempty, then we have*

$$\bigcap_\alpha K_\alpha \neq \varnothing.$$

In particular, if each K_n is nonempty and $K_{n+1} \subseteq K_n$ for every $n \in \mathbb{N}$, then we have

$$\bigcap_{n=1}^\infty K_n \neq \varnothing.$$

4.1.6   The Heine-Borel Theorem

A set E of a metric space X is said to be **bounded** if there exists a positive number M and a (fixed) point $q \in X$ such that

$$d(p, q) \leq M$$

for all $p \in E$.

In the previous subsection, we consider compact sets in an arbitrary metric space. If we take the metric space to be the Euclidean space $\mathbb{R}^n$, then there is a powerful and nice result for testing the compactness of a set. This is stated as follows:[a]

The Heine-Borel Theorem. *A subset $E \subseteq \mathbb{R}^n$ is compact if and only if E is closed and bounded in $\mathbb{R}^n$.*

---

[a]For a proof of it, please refer to [13, Theorem 2.41, p. 40].

4.1.7 Connected Sets

Definition 4.15. *Let X be a metric space. A **separation** of X is a pair E, F of disjoint nonempty open sets in X such that $X = U \cup V$. We call X **connected** if there is no separation of X.*

If the metric space X is $\mathbb{R}$, then its connected subsets have a nice structure:

Theorem 4.16. *The set $E \subseteq \mathbb{R}$ is connected if and only if for any $x, y \in E$ with $x < z < y$, we have $z \in E$.*

4.2 Open Sets and Closed Sets

Problem 4.1

$(\star)$ Let E be an open subset of X and $p \in E$. Prove that $E \setminus \{p\}$ is open.

Proof. Given $x \in E \setminus \{p\}$. Since $x \neq p$, we have $d(x, p) > 0$. Since E is open, we know that $N_r(x) \subseteq E$ for some $r > 0$. Take $r' = \min(r, \frac{d(x,p)}{2}) > 0$. If $p \in N_{r'}(x)$, then the definition gives

$$d(x, p) < r' \leq \frac{d(x, p)}{2},$$

a contradiction. Thus we have $N_{r'}(x) \subseteq E \setminus \{p\}$. Hence $E \setminus \{p\}$ is open and we finish the proof of the problem. $\blacksquare$

Problem 4.2

$(\star)$ Let E be a closed set of X and p be a limit point of E. Is the set $E \setminus \{p\}$ closed?

Proof. Since p is a limit point of E, every neighborhood $N_r(p)$ contains a point $q \neq p$ such that $q \in E$. Thus q must be an element of $E \setminus \{p\}$ and then the point p is also a limit point of $E \setminus \{p\}$. However, it is clear that $E \setminus \{p\}$ does not contain p and hence Definition 4.5 shows that it is not closed. This completes the proof of the problem. $\blacksquare$

Problem 4.3

$(\star)$ Suppose that $x, y \in X$ and $x \neq y$. Prove that there exist neighborhoods $N_r(x)$ and $N_R(y)$ of x and y respectively such that

$$N_r(x) \cap N_R(y) = \varnothing.$$

Proof. Since $x \neq y$, we have $d(x, y) > 0$. Let $\delta = \frac{d(x,y)}{4}$. Consider

$$N_\delta(x) = \{p \in X \mid d(x, p) < \delta)\} \quad \text{and} \quad N_\delta(y) = \{p \in X \mid d(y, p) < \delta)\}.$$

We claim that

$$N_\delta(x) \cap N_\delta(y) = \varnothing. \tag{4.1}$$

Otherwise, there exists a $p_0 \in N_\delta(x) \cap N_\delta(y)$. Since $p_0 \in N_\delta(x)$ and $p_0 \in N_\delta(y)$, we have

$$d(x, p_0) < \delta \quad \text{and} \quad d(y, p_0) < \delta. \tag{4.2}$$

By the triangle inequality and the inequalities (4.2), we obtain

$$d(x, y) \le d(x, p_0) + d(y, p_0) < \delta + \delta = 2\delta = \frac{d(x, y)}{2}$$

which is a contradiction. Hence we get the claim (4.1). This ends the proof of the problem. ∎

Problem 4.4

⋆ *Suppose that E is open and F is closed. Prove that $E \setminus F$ is open and $F \setminus E$ is closed.*

Proof. Recall from Problem 1.2 that $E \setminus F = E \cap F^c$. Since F is closed, F^c is open by Theorem 4.6. By Theorem 4.7(b), $E \cap F^c$ is open. Similarly, we have $F \setminus E = F \cap E^c$ and Theorem 4.7(b) implies that $F \cap E^c$ is closed. Hence we have completed the proof of the problem. ∎

Problem 4.5

⋆ *Let E and F be subsets of $\mathbb{R}$. Define*

$$E + F = \{x + y \,|\, x \in E \text{ and } y \in F\}.$$

Prove that $E + F$ is open in X if either E or F is open in X.

Proof. We just prove the case when E is open because the other case is similar. For each $y \in \mathbb{R}$, since E is open in X, $E + \{y\}$ is also open in X.[b] By definition, we have

$$E + F = \bigcup_{y \in F} (E + \{y\}).$$

By Theorem 4.7(a), $E + F$ is open in X which completes the proof of the problem. ∎

Problem 4.6

⋆ ⋆ *Prove that E is open in X if and only if E is an union of neighborhoods.*

Proof. For every $p \in E$, since E is open, there exists $r_p > 0$ such that

$$N_{r_p}(p) \subseteq E. \tag{4.3}$$

---

[b]You can imagine that the set $E + \{y\}$ is a "translation" of the set E, so every neighborhood of E is translated y units to become a neighborhood of $E + \{y\}$.

It is clear that

$$\{p\} \subseteq N_{r_p}(p) \tag{4.4}$$

for every $p \in E$. Combining the set relations (4.3) and (4.4), we have the relations

$$E = \bigcup_{p \in E} \{p\} \subseteq \bigcup_{p \in E} N_{r_p}(p) \subseteq \bigcup_{p \in E} E = E.$$

In other words, we have

$$E = \bigcup_{p \in E} N_{r_p}(p)$$

which proves the sufficient part of the problem.

Next, for the necessary part of the problem, we suppose that

$$E = \bigcup_\alpha N_\alpha,$$

where each N_α is a neighborhood of some point of X. By Theorem 4.4, each N_α is open in X and we deduce from Theorem 4.7(a) that E is also open in X. Hence we complete the proof of the problem. ∎

Problem 4.7

⊛ ⊛ Let E be closed in X. Prove that it is a countable intersection of open sets in X.

Proof. For each $n \in \mathbb{N}$ and $p \in E$, it is clear that $N_{\frac{1}{n}}(p)$ is open in X by Theorem 4.4. Therefore, Theorem 4.7(a) ensures that

$$F_n = \bigcup_{p \in E} N_{\frac{1}{n}}(p)$$

is also open in X. Now we note from the definition that $E \subseteq F_n$ for each $n \in \mathbb{N}$, so

$$E \subseteq \bigcap_{n=1}^\infty F_n. \tag{4.5}$$

Let $q \in \left(\bigcap_{n=1}^\infty F_n \right) \setminus E$. Then we have $q \in E^c$. Since E is closed in X, E^c is open in X by Theorem 4.6. Thus there is a $r > 0$ such that $N_r(q) \subseteq E^c$. By Theorem 2.1 (the Archimedean Property), one can find $m \in \mathbb{N}$ such that $mr > 1$, i.e., $\frac{1}{m} < r$. Recall that $q \in F_n$ for all $n \in \mathbb{N}$, so we fix $n = m$ to get $q \in F_m$ and thus $q \in N_{\frac{1}{m}}(p)$ *for some* $p \in E$. By definition, we have

$$d(p, q) < \frac{1}{m} < r$$

so that $p \in N_{\frac{1}{m}}(q) \subseteq N_r(q) \subseteq E^c$, but this implies that $E \cap E^c \neq \varnothing$, a contradiction. Hence we have the equality in the set relation (4.5), completing the proof of the problem. ∎

> **Problem 4.8**
>
> $\star$ Suppose that $\mathcal{V} = \{V_\alpha\}$ is a family of nonempty disjoint open subsets of $\mathbb{R}$. If $\mathcal{V}$ is infinite, prove that $\mathcal{V}$ is countable.

Proof. Let $V_\alpha \in \mathcal{V}$. Pick an arbitrary point $p_\alpha \in V_\alpha$. Since V_α is open in $\mathbb{R}$, there exists a $\epsilon_\alpha > 0$ such that

$$(p_\alpha - \epsilon_\alpha, p_\alpha + \epsilon_\alpha) \subseteq V_\alpha.$$

By Theorem 2.2 (Density of Rationals), the interval $(p_\alpha - \epsilon_\alpha, p_\alpha + \epsilon_\alpha)$ contains a rational number r_α. Take $V_\beta \in \mathcal{V}$, where $\beta \neq \alpha$. Repeat the above analysis, we can get a rational number r_β in V_β. Now $r_\alpha \neq r_\beta$ because $V_\alpha \cap V_\beta = \varnothing$. Thus what we have shown is that we take a rational number r_α from each V_α as a representative and different open subsets gives us different rational representatives. Hence the collection $\{r_\alpha\}$ is a subset of $\mathbb{Q}$. Since $\mathcal{V}$ is infinite, the set $\{r_\alpha\}$ must be infinite too. By Theorem 3.4, $\{r_\alpha\}$ must be countable, finishing our proof of the problem. $\blacksquare$

> **Problem 4.9**
>
> $\star$ $\star$ What sets are both open and closed in $\mathbb{R}$?

Proof. Suppose that E is both open and closed in $\mathbb{R}$. By Theorem 4.6, E^c is also open and closed in $\mathbb{R}$. Assume that

$$E \neq \varnothing \quad \text{and} \quad E^c \neq \varnothing.$$

Then we pick $x \in E$ and $y \in E^c$. Without loss of generality, we may assume that $x < y$. Consider the set

$$F = \{a \in \mathbb{R} \mid [x, a] \subseteq E\}.$$

Since E is open in $\mathbb{R}$, we have $(x - \epsilon, x + \epsilon) \subseteq E$ for some $\epsilon > 0$. This means $x + \frac{\epsilon}{2} \in F$ and F is nonempty. If F is not bounded above, then we have $[x, n] \subseteq E$ for infinitely many positive integers n. Take $n \geq y$ so that

$$[x, y] \subseteq [x, n] \subseteq E$$

which gives a contradiction that $y \in E$. Thus F must be bounded above. By the Completeness Axiom, we know that $\alpha = \sup F$ exists in $\mathbb{R}$.

If $\alpha \in E$, then since E is open in $\mathbb{R}$, $\alpha + \delta \in E$ for some $\delta > 0$. Thus $[x, \alpha + \delta] \subseteq E$ and then $\alpha + \delta \in F$. However, this implies the contradiction that

$$\alpha + \delta \leq \alpha.$$

Suppose that $\alpha \in E^c$. On the one hand, since E^c is open in $\mathbb{R}$, we have $\alpha - \delta' \in E^c$ for some $\delta' > 0$. Since $\alpha - \delta' < \alpha$, we follow from the definition of supremum that

$$\alpha - \delta' \leq a$$

for some $a \in F$. If $\alpha - \delta' \leq x$, then

$$[x, y] \subseteq [\alpha - \delta', y] \subseteq E^c$$

which implies that $x \in E^c$, a contradiction. Therefore, we have $x < \alpha - \delta'$ and this fact yields

$$[x, \alpha - \delta'] \subseteq [x, a] \subseteq E. \tag{4.6}$$

On the other hand, the fact $\alpha - \delta' \in E^c$ implies that $\alpha - \delta' \notin E$ and so $[x, \alpha - \delta'] \nsubseteq E$ which contradicts the set relation (4.6). Hence we obtain either $E = \varnothing$ or $E^c = \varnothing$ and each case implies that $\varnothing$ and $\mathbb{R}$ are the *only* sets that are both open and closed in $\mathbb{R}$. This completes the proof of the problem. ∎

Problem 4.10

⋆ ⋆ Let E be a subset of X and $x \in X$. Define $\rho_E(x) = \inf\{d(x, p) \mid p \in E\}$. Prove that $x \in \overline{E}$ if and only if $\rho_E(x) = 0$.

Proof. It is obvious that $\rho_E(x) \geq 0$. Suppose that $x \in \overline{E}$. Then $x \in E$ or $x \in E'$. If $x \in E$, then it is clear that $\rho_E(x) = 0$. If $x \in E'$, then it is a limit point of E. Thus *for every* $\epsilon > 0$, there is a point

$$p \in N_\epsilon(x) \cap (E \setminus \{x\})$$

In other words, we have $d(x, p) < \epsilon$ *for every* $\epsilon > 0$ and by definition, $\rho_E(x) = 0$.

Conversely, we suppose that $\rho_E(x) = 0$. If $x \in E$, then we are done. Now without loss of generality, we may assume that $x \notin E$. We claim that for every $\epsilon > 0$, there exists a $p \in E$ such that $d(x, p) < \epsilon$. Otherwise, there is a positive integer N such that

$$\frac{1}{N} \leq d(x, p)$$

for all $p \in E$ but this means that $\frac{1}{N}$ is a lower bound of the set $A_x = \{d(x, p) \mid p \in E\}$. Thus we have

$$\rho_E(x) \geq \frac{1}{N} > 0,$$

a contradiction. This proves our claim. In addition, it is easy to see that $p \neq x$ because $x \notin E$. In other words, what we have shown is that

$$N_\epsilon(x) \cap (E \setminus \{x\}) \neq \varnothing$$

for every $\epsilon > 0$. By Definition 4.8, we have $x \in E' \subseteq \overline{E}$, completing the proof of the problem. ∎

Problem 4.11

⋆ ⋆ Suppose that E° is the interior of E.

(a) Prove that E° is an open set in X.

(b) Prove that E is open in X if and only if $E^\circ = E$.

(c) Prove that if $F \subseteq E$ and F is open in X, then $F \subseteq E^\circ$.

Proof.

(a) Let $x \in E^\circ$. By Definition 4.8, x is an interior point of E and then there exists a $\epsilon > 0$ such that $N_\epsilon(x) \subseteq E$. If $y \in N_\epsilon(x)$, then Theorem 4.4 implies that

$$N_{\epsilon'}(y) \subseteq N_\epsilon(x) \subseteq E$$

for some $\epsilon' > 0$. By Definition 4.8, $y \in E^\circ$. Since y is arbitrary, we have $N_\epsilon(x) \subseteq E^\circ$ and hence E° is open in X.

(b) If $E^\circ = E$, then part (a) implies that E is open in X. Suppose that E is open in X. Recall that we always have $E^\circ \subseteq E$. Since E is open in X, every $x \in E$ is an interior point of E by Definition 4.3. Thus $x \in E^\circ$, i.e., $E \subseteq E^\circ$. Hence we have $E^\circ = E$.

(c) Let $\mathcal{F}$ be the union of all open subsets $F \subseteq E$ of X. Since $E^\circ \subseteq E$, we have

$$E^\circ \subseteq \mathcal{F}. \tag{4.7}$$

By Theorem 4.7(a), $\mathcal{F}$ is open in X. Thus for every $x \in \mathcal{F}$, there exists a $\delta > 0$ such that

$$N_\delta(x) \subseteq \mathcal{F}. \tag{4.8}$$

Since $\mathcal{F} \subseteq E$, we obtain from the set relation (4.8) that

$$N_\delta(x) \subseteq E$$

and thus $x \in E^\circ$, i.e., $\mathcal{F} \subseteq E^\circ$. Combining this and the set relation (4.7), we have $E^\circ = \mathcal{F}$. This completes the proof of the problem. ∎

Problem 4.12

$(\star)(\star)$ *For any E and F, show that*

$$E^\circ \cap F^\circ = (E \cap F)^\circ \quad \text{and} \quad E^\circ \cup F^\circ \subseteq (E \cup F)^\circ.$$

Find an example to show that the inclusion of the second assertion can be proper.

Proof. Since $E^\circ \subseteq E$ and $F^\circ \subseteq F$, we have

$$E^\circ \cap F^\circ \subseteq E \cap F \quad \text{and} \quad E^\circ \cup F^\circ \subseteq E \cup F. \tag{4.9}$$

By Problem 4.11(a), both E° and F° are open in X so we deduce from Theorem 4.7(b) that $E^\circ \cap F^\circ$ and $E^\circ \cup F^\circ$ are open in X. Now Problem 4.10(c) shows that $(E \cap F)^\circ$ and $(E \cup F)^\circ$ are the largest open subsets of X inside $E \cap F$ and $E \cup F$ respectively. This fact and the set relations (4.9) implies that

$$E^\circ \cap F^\circ \subseteq (E \cap F)^\circ \quad \text{and} \quad E^\circ \cup F^\circ \subseteq (E \cup F)^\circ. \tag{4.10}$$

Thus the second set relation in (4.10) is one of our expected results.

Since $(E \cap F)^\circ \subseteq E \cap F$, we acquire that

$$(E \cap F)^\circ \subseteq E \quad \text{and} \quad (E \cap F)^\circ \subseteq F. \tag{4.11}$$

Since E° and F° are the largest open subsets of X inside E and F respectively, it follows from this and the set relations (4.11) that

$$(E \cap F)^\circ \subseteq E^\circ \subseteq E \quad \text{and} \quad (E \cap F)^\circ \subseteq F^\circ \subseteq F.$$

Therefore, we have

$$(E \cap F)^\circ \subseteq E^\circ \cap F^\circ. \tag{4.12}$$

Combining the first set relation in (4.10) and the set relation (4.12), we obtain the remaining expected result that

$$(E \cap F)^\circ = E^\circ \cap F^\circ.$$

Consider the sets $E = [0, 1]$ and $F = [1, 2]$. Then we have

$$E^\circ = (0, 1), \quad F^\circ = (1, 2) \quad \text{and} \quad (E \cup F)^\circ = (0, 2).$$

In this case, we have $E^\circ \cup F^\circ \subset (E \cup F)^\circ$, completing the proof of the problem. ∎

Problem 4.13

⋆ ⋆ Let E and F be subsets of X.

(a) Show that $\overline{E \cap F} \subseteq \overline{E} \cap \overline{F}$.

(b) Show that $E \cap \overline{F} \subseteq \overline{E \cap F}$ if E is open.

(c) Find an example to show that the inclusion in parts (a) and (b) can be proper.

Proof.

(a) Note that

$$E \cap F \subseteq E \subseteq \overline{E} \tag{4.13}$$

and $\overline{E}$ is closed in X by Theorem 4.9(a). Since $\overline{E \cap F}$ is the smallest closed subset of X containing $E \cap F$, this and the set relation (4.13) implies that $\overline{E \cap F} \subseteq \overline{E}$. Similarly, $\overline{E \cap F} \subseteq \overline{F}$ also holds and so

$$\overline{E \cap F} \subseteq \overline{E} \cap \overline{F}.$$

(b) Suppose that E is open in X. Given $\epsilon > 0$. If $x \in E \cap \overline{F}$, then $x \in E$ and so *there exists* a $\delta > 0$ such that $N_\delta(x) \subseteq E$. Since $x \in \overline{F}$, x is a limit point of F. Take $\epsilon' = \min(\epsilon, \delta)$. By Definition 4.5, we have

$$N_{\epsilon'}(x) \cap (F \setminus \{x\}) \neq \varnothing.$$

Let $y \in N_{\epsilon'}(x) \cap (F \setminus \{x\})$. Since $N_{\epsilon'}(x) \subseteq N_\delta(x) \subseteq E$, we always have

$$y \in N_{\epsilon'}(x) \cap (F \setminus \{x\}) \subseteq E \cap (F \setminus \{x\}). \tag{4.14}$$

Furthermore, $y \in N_{\epsilon'}(x) \subseteq N_\epsilon(x)$. It follows from this and the set relation (4.14) that

$$N_\epsilon(x) \cap [(E \cap F) \setminus \{x\}] \neq \varnothing$$

for every $\epsilon > 0$. By Definition 4.5, $x \in \overline{E \cap F}$ and hence $E \cap \overline{F} \subseteq \overline{E \cap F}$.

(c) Take $E = (0, 1)$ and $F = (1, 2)$. Then it is easy to check that

$$\overline{E \cap F} = \overline{\varnothing} = \varnothing \quad \text{and} \quad \overline{E} \cap \overline{F} = \{1\},$$

so the inclusion in parts (a) and (b) can be proper.

This completes the proof of the problem. ∎

Problem 4.14

(⋆) Prove that $E^\circ = X \setminus (\overline{X \setminus E})$.

Proof. Now $x \in E^\circ$ if and only if *there exists* a $\epsilon > 0$ such that $N_\epsilon(x) \subseteq E$ (by Definition 4.3) if and only if $N_\epsilon(x) \cap (X \setminus E) = \varnothing$ (by difference of sets in §1.1.1) if and only if $x \notin \overline{X \setminus E}$ (by Definition 4.5) if and only if $x \in X \setminus (\overline{X \setminus E})$ (by difference of sets in §1.1.1). Hence we have the desired formula and we complete the proof of the problem. ∎

Problem 4.15

(⋆) Find an example to show that the inclusion in Theorem 4.11(b) can be proper.

Proof. Consider $X = \mathbb{R}$, $Y = \mathbb{Q}$ and $E = Y \cap (0, 1)$. On the one hand, since $(0, 1)$ is open in $\mathbb{R}$, E is open in Y by Theorem 4.10. By Problem 4.11(b), we know that

$$E_Y^\circ = E = Y \cap (0, 1). \tag{4.15}$$

On the other hand, if $x \in Y_X^\circ$, then x is an interior point of Y and thus there exists a $\epsilon > 0$ such that $(x - \epsilon, x + \epsilon) \subseteq Y$. However, Problem 2.9 guarantees that the existence of an irrational number θ lying in $(x - \epsilon, x + \epsilon)$ and this implies that $\theta \in Y$, a contradiction. Hence we must have

$$Y_X^\circ = \varnothing. \tag{4.16}$$

Since $E \subseteq Y$ implies that $E^\circ \subseteq Y^{\circ},$[c] we obtain from this and the fact (4.16) that

$$E_X^\circ \subseteq Y_X^\circ = \varnothing.$$

In other words, we have $E_X^\circ \cap Y = \varnothing$ which is clearly a proper subset of the set (4.15). This ends the proof of the problem. ∎

4.3 Compact Sets

Problem 4.16

(⋆)(⋆) Suppose that $K \subseteq U \cup V$, where U and V are disjoint open subsets of X and K is compact. Prove that $K \cap U$ is compact.

---

[c]See Table 4.1.

Proof. Since $U \cap V = \varnothing$, we have $U \subseteq V^c$. Since $U \cup U^c = X$, we have

$$
\begin{aligned}
K \cap V^c &= (K \cap V^c) \cap X \\
&= (K \cap V^c) \cap (U \cup U^c) \\
&= [(K \cap V^c) \cap U] \cup [(K \cap V^c) \cap U^c].
\end{aligned} \tag{4.17}
$$

Let $x \in K$. Then $x \in U \cup V$ so that $x \notin U^c \cap V^c$. In other words, $K \cap (U^c \cap V^c) = \varnothing$. Thus we put this into (4.17) to get

$$
K \cap V^c = (K \cap U) \cup \varnothing = K \cap U. \tag{4.18}
$$

Since V^c is closed in X, Theorem 4.13(b) implies that $K \cap V^c$ is compact. By the representation (4.18), we conclude that $K \cap U$ is also compact. This finishes the proof of the problem. ∎

Problem 4.17

⋆ ⋆ *Suppose that $K_1, \ldots, K_n$ are compact sets. Prove that $K = K_1 \cup K_2 \cup \cdots \cup K_n$ is also compact.*

Proof. Let $\{V_\alpha\}$ be an open cover of K. Since K_i is compact, there exists $V_{\alpha_{i_1}}, V_{\alpha_{i_2}}, \ldots, V_{\alpha_{i_m}}$ such that

$$
K_i \subseteq V_{\alpha_{i_1}} \cup V_{\alpha_{i_2}} \cup \cdots \cup V_{\alpha_{i_m}} = \bigcup_{j=1}^{m} V_{\alpha_{i_j}}
$$

which implies that

$$
K = \bigcup_{i=1}^{n} K_i \subseteq \bigcup_{i=1}^{n} \bigcup_{j=1}^{m} V_{\alpha_{i_j}}
$$

By Definition 4.12 (Compact Sets), K is compact and this completes the proof of the problem. ∎

Problem 4.18

⋆ ⋆ *Suppose that $\{K_\alpha\}$ is a collection of compact sets. Prove that*

$$
K = \bigcap_\alpha K_\alpha
$$

is also compact.

Proof. Since each K_α is compact, it is closed in X by Theorem 4.13(a). By Theorem 4.7(a), the set K must be closed in X. It is clear that

$$
K = K \cap K_\alpha
$$

for every α. Now an application of Theorem 4.13(b) shows that K is compact and we finish the proof of the problem. ∎

Problem 4.19

$\bigstar$ $\bigstar$ *Prove that the closed interval $[0, 1]$ is compact from the definition.*

Proof. Let $\{V_\alpha\}$ be an open cover of $[0, 1]$. Consider the set

$$E = \{x \in [0, 1] \mid [0, x] \text{ can be covered by finitely many } V_\alpha\}.$$

It is clear that $E \neq \varnothing$ because $0 \in E$. Furthermore, it is also bounded above by 1. By the Completeness Axiom, we know that $\beta = \sup E$ exists in $\mathbb{R}$. Since $0 \in V_\alpha$ for some α and V_α is open in $\mathbb{R}$, we have $(-\eta, \eta) \subseteq V_\alpha$ for some $\eta > 0$. Then it implies that $[0, \frac{\eta}{2}] \subseteq V_\alpha$ and this means that $\frac{\eta}{2} \in E$. By this fact and the definition of supremum, we must have

$$\beta > 0. \tag{4.19}$$

Assume that $\beta < 1$. Since $\{V_\alpha\}$ covers $[0, 1]$, $\beta \in V_{\alpha_0}$ for some α_0. Since V_{α_0} is open in $\mathbb{R}$, there exists a $\delta > 0$ such that

$$\beta \in (\beta - \delta, \beta + \delta) \subseteq V_{\alpha_0}. \tag{4.20}$$

We claim that $[0, \beta]$ can be covered by finitely many V_α. By the fact (4.19) and Theorem 2.2 (Density of Rationals), we can choose δ so small that $\beta - \delta > 0$. Since $\beta - \frac{\delta}{2} < \beta$, there exists a $x \in E$ such that $\beta - \frac{\delta}{2} \leq x < \beta$. Since $[0, \beta - \frac{\delta}{2}] \subseteq [0, x]$, the interval $[0, \beta - \frac{\delta}{2}]$ is covered by finitely many V_α. Therefore, we follow from the these, the set relation (4.20) and the fact $[0, \beta] \subseteq [0, \beta - \frac{\delta}{2}] \cup (\beta - \delta, \beta]$ that $[0, \beta]$ is covered by finitely many V_α. This prove the claim, i.e.,

$$[0, \beta] \subseteq V_{\alpha_1} \cup V_{\alpha_2} \cup \cdots \cup V_{\alpha_m}. \tag{4.21}$$

Furthermore, it follows from the set relations (4.20) and (4.21) that

$$\begin{aligned}
\left[0, \beta + \frac{\delta}{2}\right] &= [0, \beta] \cup \left[\beta, \beta + \frac{\delta}{2}\right] \\
&\subseteq V_{\alpha_1} \cup V_{\alpha_2} \cup \cdots \cup V_{\alpha_m} \cup (\beta - \delta, \beta + \delta) \\
&\subseteq V_{\alpha_1} \cup V_{\alpha_2} \cup \cdots \cup V_{\alpha_m} \cup V_{\alpha_0}
\end{aligned}$$

which implies that $\beta + \frac{\delta}{2} \in E$, but this certainly contradicts to the fact that $\beta = \sup E$. Hence we must have $\beta = 1$ and then $[0, 1]$ is compact, completing the proof of the problem. ∎

Problem 4.20

$\bigstar$ $\bigstar$ *Construct an open cover of $(0, 1)$ which has no finite subcover so that $(0, 1)$ is not compact.*

Proof. For each $n = 2, 3, \ldots$, we consider the interval $V_n = (\frac{1}{n}, 1)$. If $x \in (0, 1)$, then it follows from Theorem 2.1 (the Archimedean property) that there exists a positive integer n such that $nx > 1$, i.e., $x \in V_n$. Furthermore, we have

$$(0, 1) \subseteq \bigcup_{i=2}^{\infty} V_n,$$

i.e., $\{V_2, V_3, \ldots\}$ is an open cover of the segment $(0, 1)$.

Assume that $\{V_{n_1}, V_{n_2}, \ldots, V_{n_k}\}$ was a finite subcover of $(0, 1)$, where $n_1, n_2, \ldots, n_k$ are positive integers and $2 \le n_1 < n_2 < \cdots < n_k$. By definition, we have

$$V_{n_1} \subseteq V_{n_2} \subseteq \cdots \subseteq V_{n_k}$$

and so

$$(0, 1) \subseteq \bigcup_{i=1}^{k} V_{n_i} \subseteq V_{n_k},$$

contradicting to the fact that $\frac{1}{2n_k} \in (0, 1)$ but $\frac{1}{2n_k} \notin (\frac{1}{n_k}, 1)$. Hence $\{V_2, V_3, \ldots\}$ does not have a finite subcover for $(0, 1)$. This completes the proof of the problem. ∎

Remark 4.1

Problems 4.19 and 4.20 tell us the fact that arbitrary subsets of a compact set need not be compact.

Problem 4.21

⋆ ⋆ Suppose that $\{K_\alpha\}$ is a collection of compact subsets of X. Prove that if $\bigcap_\alpha K_\alpha = \varnothing$, then there is a choice of finitely many indices $\alpha_1, \ldots, \alpha_n$ such that

$$K_{\alpha_1} \cap K_{\alpha_2} \cap \cdots \cap K_{\alpha_n} = \varnothing.$$

Proof. Let $K \in \{K_\alpha\}$. By Theorem 4.13(a), each K_α is closed in X so that K_α^c is open in X by Theorem 4.6. For every $x \in K$, since $\bigcap_\alpha K_\alpha = \varnothing$, $x \notin K_\alpha$ for some α and this is equivalent to saying that

$$x \in K_\alpha^c.$$

Therefore, $\{K_\alpha^c\}$ is an open cover of K. Since K is compact, a finite subcover $K_{\alpha_1}^c, K_{\alpha_2}^c, \ldots, K_{\alpha_n}^c$ of K exists, i.e.,

$$K \subseteq K_{\alpha_1}^c \cup K_{\alpha_2}^c \cup \ldots \cup K_{\alpha_n}^c. \tag{4.22}$$

Applying complements to both sides in (4.22) and using Problem 1.1, we get

$$K_{\alpha_1} \cap K_{\alpha_2} \cap \cdots \cap K_{\alpha_n} \subseteq K^c$$

which implies that

$$K_{\alpha_1} \cap K_{\alpha_2} \cap \cdots \cap K_{\alpha_n} \cap K = \varnothing.$$

This ends the proof of the problem. ∎

Problem 4.22

$(\star)(\star)$ *Suppose that $K \neq \varnothing$ is a compact metric space and $\{E_n\}$ is a sequence of nonempty closed subsets of K such that $E_{n+1} \subseteq E_n$ for every $n \in \mathbb{N}$. Verify that*

$$E = \bigcap_{n=1}^{\infty} E_n \neq \varnothing.$$

Proof. Assume that $E = \varnothing$. We claim that

$$K \subseteq E^c.$$

Otherwise, there exists a $x \in K$ such that $x \notin E^c$. This means that $x \notin E_n^c$ for every $n \in \mathbb{N}$ or equivalently, $x \in E_n$ for every $n \in \mathbb{N}$. However, this implies that $x \in E$ so that $E \neq \varnothing$, a contradiction.

Recall that each E_n is closed in K, so we get from Theorem 4.6 that E_n^c is open in K and then $\{E_n^c\}$ forms an open covering of K becuase

$$E^c = \bigcup_{n=1}^{\infty} E_n^c.$$

By Definition 4.12 (Compact Sets), one can find a finitely many indices $n_1, \ldots, n_k$ such that

$$K \subseteq E_{n_1}^c \cup E_{n_2}^c \cup \cdots \cup E_{n_k}^c. \tag{4.23}$$

Let x be a point of K such that $x \in E_{n_1} \cap \cdots \cap E_{n_k}$. Then Problem 1.1 implies that

$$x \notin E_{n_1}^c \cup E_{n_2}^c \cup \cdots \cup E_{n_k}^c. \tag{4.24}$$

Combining the two facts (4.23) and (4.24), we must have $x \notin K$, but this contradicts to our hypothesis that $x \in K$. Therefore, we have

$$E_{n_1} \cap \cdots \cap E_{n_k} = \varnothing. \tag{4.25}$$

Without loss of generality, we assume that $n_1 \leq n_2 \leq \cdots \leq n_k$. Since $E_{n+1} \subseteq E_n$ for every $n \in \mathbb{N}$, we see that

$$E_{n_k} \subseteq E_{n_{k-1}} \subseteq \cdots \subseteq E_{n_2} \subseteq E_{n_1}.$$

Put this chain into the left-hand side of the expression (4.25), we obtain

$$E_{n_k} = \varnothing$$

which contradicts the hypothesis that each E_n is nonempty. Hence we must have $E \neq \varnothing$ and thus finishing the proof of the problem. ∎

Problem 4.23

⋆ ⋆ ⋆ *Let $\mathcal{F}$ be a collection of subsets of X. We say $\mathcal{F}$ has the **finite intersection property** if for every finite intersection of sets from $\mathcal{F}$, $\{F_1, F_2, \ldots, F_n\} \subseteq \mathcal{F}$, we have*

$$\bigcap_{k=1}^{n} F_k \neq \varnothing.$$

Prove that a metric space K is compact if and only if every collection of closed sets $\mathcal{F}$ with the finite intersection property has a nonempty intersection.

Proof. Let $\{F_\alpha\}$ be a collection of closed sets with the finite intersection property. Suppose that the metric space K is compact. If $\bigcap_\alpha F_\alpha = \varnothing$, then we have

$$K = \bigcup_\alpha F_\alpha^c. \tag{4.26}$$

Since each F_α is closed in K, we know from Theorem 4.6 that F_α^c is open in K. Therefore, it is easy to see from the expression (4.26) that $\{F_\alpha^c\}$ is an open cover of K. By the compactness of K, a choice of finitely many indices $\alpha_1, \ldots, \alpha_n$ exists so that

$$K = F_{\alpha_1}^c \cup \cdots \cup F_{\alpha_n}^c. \tag{4.27}$$

By Problem 1.1, this fact (4.27) implies that

$$F_{\alpha_1} \cap \cdots \cap F_{\alpha_n} = (F_{\alpha_1}^c \cup \cdots \cup F_{\alpha_n}^c)^c = K^c = \varnothing$$

which contradicts the basic assumption that $\{F_\alpha\}$ has the finite intersection property. Therefore, we conclude that

$$\bigcap_\alpha F_\alpha \neq \varnothing. \tag{4.28}$$

Conversely, let $\{V_\beta\}$ be an open covering of K. Note that each V_β^c must be closed in K by Theorem 4.6 and thus $\{V_\beta^c\}$ forms a collection of closed subsets of K. Assume that $K \neq \bigcup_{k=1}^{n} V_{\beta_k}$ for *all* finite indices $\beta_1, \ldots, \beta_n$. By this and Problem 1.1, we know that

$$\bigcap_{k=1}^{n} V_{\beta_k}^c = \Big(\bigcup_{k=1}^{n} V_{\beta_k} \Big)^c \neq \varnothing.$$

Therefore, it means that $\{V_\beta^c\}$ satisfies the finite intersection property. Hence we follow from the hypothesis that the condition (4.28) holds for $\{V_\beta^c\}$. Finally, Problem 1.1 again implies that

$$\varnothing \neq \bigcap_\beta V_\beta^c = \Big(\bigcup_\beta V_\beta \Big)^c$$

or equivalently,

$$\bigcup_\beta V_\beta \neq K$$

which definitely contradicts to the fact that $\{V_\beta\}$ is an open cover of K. Hence $\{V_\beta\}$ must have a finite subcover for K and then K is compact. We have completed the proof of the problem. ■

4.4 The Heine-Borel Theorem

Problem 4.24

(⋆) *Prove that the sets* $\mathbb{Q}$, $\mathbb{R}$ *and* $E = \{\frac{1}{n} \mid n = 1, 2, \ldots\}$ *are not compact.*

Proof. The first two sets are not compact because they are not bounded. Since 0 is a limit point of E but $0 \notin E$, E is not closed. By the Heine-Borel Theorem, E is not compact. We end the proof of the problem. ∎

Problem 4.25

(⋆) *Prove that* $[0, 1]$ *is compact but* $(0, 1)$ *is not by using the Heine-Borel Theorem.*

Proof. Since $[0, 1]$ is closed and bounded in $\mathbb{R}$, it is compact by the Heine-Borel Theorem. Since $(0, 1)$ is not closed in $\mathbb{R}$, it is not compact by the Heine-Borel Theorem. This ends the proof of the problem. ∎

Remark 4.2

If you compare Problem 4.25 with Problems 4.19 and 4.20, you will see how powerful the Heine-Borel Theorem is.

Problem 4.26

(⋆) *Prove that every finite subset of* $\mathbb{R}$ *is compact.*

Proof. Let $E = \{x_1, \ldots, x_n\}$ be a finite subset of $\mathbb{R}$. Without lost of generality, we may assume that $x_1 \le x_2 \le \cdots \le x_n$. Then it is easy to see that E is bounded by $\max(|x_1|, |x_n|)$. For each x_k ($k = 1, 2, \ldots, n$), since the set

$$\mathbb{R} \setminus \{x_k\} = (-\infty, x_k) \cup (x_k, \infty)$$

is open in $\mathbb{R}$, the point set $\{x_k\}$ is closed in $\mathbb{R}$. By Theorem 4.7(b), E is also closed in $\mathbb{R}$. Hence it follows from the Heine-Borel Theorem that E is compact. This ends the proof of the problem. ∎

Problem 4.27

(⋆)(⋆) *Let* K *be a nonempty compact set in* $\mathbb{R}$. *Prove that* $\sup K, \inf K \in K$.

Proof. By the Heine-Borel Theorem, K must be bounded. Besides, K is nonempty. Thus, by the Completeness Axiom, we conclude that $\sup K$ and $\inf K$ exist in $\mathbb{R}$. Let $\beta = \sup K$ and $\alpha = \inf E$.

Assume that $\beta \notin K$. Since K is compact, it is closed in $\mathbb{R}$ by Theorem 4.13(a). Then β is *not* a limit point of K, i.e., there exists a $\epsilon > 0$ such that

$$(\beta - \epsilon, \beta + \epsilon) \cap K = \varnothing. \tag{4.29}$$

In other words, we have

$$(\beta - \epsilon, \beta + \epsilon) \subseteq \mathbb{R} \setminus K.$$

If $\beta - \frac{\epsilon}{2}$ is not an upper bound of K, then the definition of supremum shows that there is a $x \in K$ such that $\beta - \frac{\epsilon}{2} \leq x$ which implies that

$$x \in \left[\beta - \frac{\epsilon}{2}, \beta + \frac{\epsilon}{2}\right] \subset (\beta - \epsilon, \beta + \epsilon),$$

but this definitely contradicts the result (4.29). Then $\beta - \frac{\epsilon}{2}$ is an upper bound of K which is a contradiction. Hence we conclude that $\beta \in K$. The case for proving $\alpha \in K$ is very similar, so we omit the details here. Now we complete the proof of the problem. ∎

Problem 4.28

($\star$) *Suppose that $E_0 = [0,1]$. Let E_1 be the set obtained from E_0 by removing the "middle third" $\left(\frac{1}{3}, \frac{2}{3}\right)$. Next, let E_2 be the set obtained from E_1 by removing the "middle thirds" $\left(\frac{1}{9}, \frac{2}{9}\right)$ and $\left(\frac{7}{9}, \frac{8}{9}\right)$. In fact, E_n is given by*

$$E_n = \bigcup_{k=0}^{2^{n-1}-1} \left(\left[\frac{3k+0}{3^n}, \frac{3k+1}{3^n}\right] \cup \left[\frac{3k+2}{3^n}, \frac{3k+3}{3^n}\right]\right), \tag{4.30}$$

where $n \geq 1$. The intersection

$$C = \bigcap_{n=1}^{\infty} E_n \tag{4.31}$$

*is the well-known **Cantor set**. Prove that C is compact.*

Proof. It is easy to see from the definition (4.30) that each E_n is a union of 2^n closed intervals in $\mathbb{R}$, so Theorem 4.7(b) implies that E_n is closed in $\mathbb{R}$. Next, we see from the definition (4.31) that C is the intersection of closed subsets in $\mathbb{R}$, so it is also closed in $\mathbb{R}$ by Theorem 4.7(a). It is clear that C is a bounded set. Hence we deduce from the Heine-Borel Theorem that C is compact, finishing the proof of the problem. ∎

4.5 Connected Sets

Problem 4.29

($\star$) *Show that $\mathbb{Q}$ is not connected.*

Proof. Let $a, b \in \mathbb{Q}$ with $a < b$. By Problem 2.9, we see that there is an irrational number θ such that $a < \theta < b$. Since $\theta \notin \mathbb{Q}$, it follows from Theorem 4.16 that $\mathbb{Q}$ is not connected. This completes the proof of the problem. ∎

Problem 4.30

⋆ ⋆ *Suppose that the pair U and V forms a separation of X and E is a connected subset of X. Prove that E is contained either in U or V.*

Proof. Assume that $E \cap U \neq \varnothing$ and $E \cap V \neq \varnothing$. By Theorem 4.10, we deduce that both $E \cap U$ and $E \cap V$ are open in E. Since U and V are disjoint, it is easy to see that

$$(E \cap U) \cap (E \cap V) = E \cap (U \cap V) = E \cap \varnothing = \varnothing$$

so that $E \cap U$ and $E \cap V$ are also disjoint. Furthermore, we have

$$(E \cap U) \cup (E \cap V) = E.$$

By Definition 4.15, $E \cap U$ and $E \cap V$ form a separation of E which is a contradiction. Hence we have either

$$E \subseteq U \quad \text{or} \quad E \subseteq V.$$

We complete the proof of the problem. ∎

Problem 4.31

⋆ ⋆ *Suppose that $\{E_\alpha\}$ is a family of connected subsets of X and $\bigcap_\alpha E_\alpha \neq \varnothing$. Prove that*

$$E = \bigcup_\alpha E_\alpha$$

is connected.

Proof. Let $p \in \bigcap_\alpha E_\alpha$. Assume that E had a separation, i.e., $E = U \cup V$, where U and V are disjoint nonempty open subsets of X. Now we have either $p \in U$ or $p \in V$. Let $p \in U$. Since each E_α is connected, it must be contained in either U or V by Problem 4.30. Since $p \in E_\alpha$ for every α and $p \in U$, we have $E_\alpha \subseteq U$ for every α which implies that $E \subseteq U$ or equivalently $V = \varnothing$. This is clearly a contradiction, so we have the desired result that E is connected. We complete the proof of the problem. ∎

Problem 4.32

⋆ ⋆ *Let U and V form a separation of E. Prove that*

$$\overline{U} \cap V = U \cap \overline{V} = \varnothing.$$

Proof. By Definition 4.15, U and V are disjoint nonempty open sets in X and $X = U \cup V$. Since $U^c = X \setminus U = V$, U is also closed in X by Theorem 4.6. Then it follows from Theorem 4.9(b) that $\overline{U} = U$. Recall that $U \cap V = \varnothing$, so we have

$$\overline{U} \cap V = U \cap V = \varnothing.$$

The case for $U \cap \overline{V} = \varnothing$ can be proven similarly, so we omit the proof here. This completes the proof of the problem. ∎

Problem 4.33

$\bigstar$ $\bigstar$ *Suppose that E is connected. Prove that $\overline{E}$ is connected.*

Proof. Assume that the pair U and V was a separation of $\overline{E}$. By Problem 4.30, we have either $E \subseteq U$ or $E \subseteq V$. Without loss of generality, we may assume that $E \subseteq U$. By the proof of Problem 4.32, we see that U is closed in $\overline{E}$. Then we establish from Theorem 4.9(c) that

$$\overline{E} \subseteq U$$

and this means that $V = \varnothing$, a contradiction. Hence $\overline{E}$ is connected and this completes the proof of the problem. ∎

Problem 4.34

$\bigstar$ $\bigstar$ *If $E \subseteq \mathbb{R}$ is connected, prove that E° is also connected.*

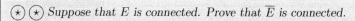

Proof. Assume that E° was not connected. By Theorem 4.16, there exist $x, y \in E^\circ$ and a number z with $x < z < y$ but $z \notin E^\circ$. Since $E^\circ \subseteq E$, we always have $x, y \in E$. Since E is connected, Theorem 4.16 implies that $(x, y) \subseteq E$. In particular, we have

$$z \in (x, y) \subseteq E. \tag{4.32}$$

By Theorem 2.2 (Density of Rationals) and the set relation (4.32), there exists a $\epsilon > 0$ such that $N_\epsilon(z) \subseteq (x, y) \subseteq E$, therefore

$$z \in E^\circ$$

which contradicts to the assumption. Hence E° is connected and we end the proof of the problem. ∎

CHAPTER 5

Sequences in Metric Spaces

▌ 5.1 Fundamental Concepts

In fundamental calculus, we have learnt the concepts of sequences of real (or complex) numbers. In fact, those concepts can be generalized and explained without difficulties in a general setting of sequences in metric spaces. The main references for this part are [3, §4.1 - §4.4, pp. 70 - 74], [5, §3.1 - §3.6], [6, Chap. 2], [13, Chap. 3] and [15, Chap. 5 and 6].

▌ 5.1.1 Convergent Sequences in Metric Spaces

Let $\{p_n\}$ be a sequence of points in a metric space X with the metric d. We say that it **converges** if there is a point $p \in X$ such that *for every* $\epsilon > 0$, *there exists* a positive integer N such that

$$d(p_n, p) < \epsilon$$

for all $n \geq N$. In notation, we write $p_n \to p$ as $n \to \infty^{\text{a}}$ or

$$\lim_{n \to \infty} p_n = p. \tag{5.1}$$

The point p in the limit (5.1) is called the **limit** of $\{p_n\}$. If $\{p_n\}$ does not converge, then we say that it is **diverge**. A sequence $\{p_n\}$ is called **bounded** if there exists a positive number M and a (fixed) point $q \in X$ such that

$$d(p_n, q) \leq M$$

for all $n \in \mathbb{N}$.

In the following, $\{a_n\}, \{b_n\}, \{p_n\}, \{q_n\}, \{x_y\}, \{y_n\}, \ldots$ will always denote sequences in a metric space X. Some basic facts about convergent sequences are given as follows:

Theorem 5.1 (Uniqueness of Limits of Sequences). *A sequence $\{p_n\}$ can converge to at most one point in X.*

Theorem 5.2. *Let $\{p_n\}$ be a sequence in X and $p_n \to p$, where $p \in X$. Then $\{p_n\}$ is bounded and p is a limit point of the set $E = \{p_1, p_2, \ldots\}$.*

---

[a]Or simply $p_n \to p$.

Given a set $\{n_1, n_2, \ldots\}$ of positive integers such that $n_1 < n_2 < \cdots$. Then the sequence $\{p_{n_k}\}$ is a **subsequence** of the original sequence $\{p_n\}$. The following result characterizes the convergence of $\{p_n\}$ and convergence of its subsequences $\{p_{n_k}\}$:

Theorem 5.3. *In a metric space X, $\{p_n\}$ converges to p if and only if every subsequence $\{p_{n_k}\}$ converges to p.*

Next, the following result can be treated as a converse to the second assertion in Theorem 5.2:

Theorem 5.4. *Suppose that $E \subseteq X$. If p is a limit point of E, then there is a sequence $\{p_n\}$ in E such that the result (5.1) holds.*

5.1.2   Sequences in $\mathbb{R}^n$ and Some Well-known Sequences

If the metric space X is taken to be $\mathbb{R}^n$ for some $n \in \mathbb{N}$, then the convergence of sequences in $\mathbb{R}^n$ and the algebraic operations of points in $\mathbb{R}^n$ can be summarized as follows:

Theorem 5.5. *Suppose that $\mathbf{p}_k \in \mathbb{R}^n$, where $k \in \mathbb{N}$ and*

$$\mathbf{p}_k = (p_{1k}, p_{2k}, \ldots, p_{nk}).$$

Then $\mathbf{p}_k \to \mathbf{p} = (p_1, \ldots, p_n)$ as $k \to \infty$ if and only if

$$\lim_{k \to \infty} p_{ik} = p_i,$$

where $i = 1, 2, \ldots, n$. Furthermore, if $\mathbf{p}_k, \mathbf{q}_k \in \mathbb{R}^n$ and $c_k \in \mathbb{R}$ for $k \in \mathbb{N}$ and $\mathbf{p}_k \to \mathbf{p}$, $\mathbf{q}_k \to \mathbf{q}$, $c_k \to c$ as $k \to \infty$, then we have

$$\lim_{k \to \infty} (\mathbf{p}_k + \mathbf{q}_k) = \mathbf{p} + \mathbf{q}, \quad \lim_{k \to \infty} (\mathbf{p}_k \cdot \mathbf{q}_k) = \mathbf{p} \cdot \mathbf{q} \quad and \quad \lim_{k \to \infty} c_k \mathbf{p}_k = c\mathbf{p}.$$

We say that a sequence $\{p_n\}$ of real numbers is **monotonically increasing** (resp. **monotonically decreasing**) if $p_n \le p_{n+1}$ (resp. $p_n \ge p_{n+1}$) for all $n \in \mathbb{N}$. A **monotonic sequence** is either a monotonically increasing or monotonically decreasing sequence. Now we have the so-called **Monotonic Convergence Theorem**:

The Monotonic Convergence Theorem. *Every monotonic and bounded sequence $\{p_n\}$ is convergent.*

Finally, the following two theorems state the **Squeeze Theorem for Convergent Sequences**[b] and some well-known limits of sequences in $\mathbb{R}$ and we are going to use them frequently in the problems and solutions.

Theorem 5.6 (Squeeze Theorem for Convergent Sequences)**.** *Suppose that $\{a_n\}$, $\{b_n\}$ and $\{c_n\}$ are sequences of real numbers and $a_n \le c_n \le b_n$ for all but finitely many positive integers n. If $\lim\limits_{n \to \infty} a_n = \lim\limits_{n \to \infty} b_n = \ell$, then we have*

$$\lim_{n \to \infty} c_n = \ell.$$

---

[b]This is also called the **Sandwich Theorem**.

Theorem 5.7.

(a) If $k > 0$, then $\lim\limits_{n \to \infty} \dfrac{1}{n^k} = 0$ and $\lim\limits_{n \to \infty} \sqrt[n]{k} = 1$.

(b) $\lim\limits_{n \to \infty} \sqrt[n]{n} = 1$.

(c) If $|x| < 1$, then $\lim\limits_{n \to \infty} x^n = 0$.

▌5.1.3 Upper Limits and Lower Limits

The **extended real number system**, denoted by $[-\infty, +\infty]$, consists of $\mathbb{R}$ and the two symbols $+\infty$ and $-\infty$.[c] Then it is obvious that $+\infty$ is an upper bound of every subset of $[-\infty, +\infty]$ and every nonempty subset of it has the least upper bound.[d] This set preserves the original operations of $\mathbb{R}$ when we have $x, y \in \mathbb{R}$. When x is real and y is either $+\infty$ or $-\infty$, then we have the following conventions:

(a) $x + (+\infty) = +\infty$, $x + (-\infty) = -\infty$, $x - (+\infty) = -\infty$ and $x - (-\infty) = +\infty$, $\dfrac{x}{\pm\infty} = 0$.

(b) If $x > 0$, then $x \cdot (+\infty) = +\infty$ and $x \cdot (-\infty) = -\infty$.

(c) If $x < 0$, then $x \cdot (+\infty) = -\infty$ and $x \cdot (-\infty) = +\infty$.

(d) $(+\infty) + (+\infty) = (+\infty) \cdot (+\infty) = (-\infty) \cdot (-\infty) = +\infty$.

(e) $(-\infty) + (-\infty) = (+\infty)(-\infty) = -\infty$.

Definition 5.8. *Suppose that $\{x_n\}$ is a sequence in $[-\infty, +\infty]$. Let E be the set of all* **subsequential limits**, *i.e., if $x \in E$, then $\{x_n\}$ has a subsequence $\{x_{n_k}\}$ such that*

$$\lim_{k \to \infty} x_{n_k} = x.$$

It is allowed that E contains $+\infty$ and $-\infty$.

By the above discussion, we know that $\sup E$ and $\inf E$ exists in $[-\infty, +\infty]$. They are called the **upper limit** and **lower limit** of $\{x_n\}$ respectively. In notation, we have

$$\limsup_{n \to \infty} x_n = \sup E \quad \text{and} \quad \liminf_{n \to \infty} x_n = \inf E. \tag{5.2}$$

The following theorem shows some important properties of $\sup E$. An analogous result is also valid for $\inf E$, so we won't repeat the statement here.

Theorem 5.9. *The upper limit $\sup E$ is the only number satisfying the following properties:*

(a) $\sup E \in E$.

---

[c]We remark that $[-\infty, +\infty]$ is *not* a field.

[d]The case for the definition of the greatest lower bound is exactly the same as the least upper bound, so we won't repeat here.

(b) If $x > \sup E$, then there exists a positive integer N such that

$$x_n < x$$

for all $n \geq N$.

Next, it is quite obvious that the following theorem holds:

Theorem 5.10. *Suppose that $\{x_n\}$ is a real sequence. Then $\lim\limits_{n\to\infty} x_n = x$ if and only if*

$$\limsup_{n\to\infty} x_n = \liminf_{n\to\infty} x_n = x.$$

Sometimes we need to compare upper limits or lower limits of two different sequences and the following result is very useful to deal with this question:

Theorem 5.11. *Suppose that N is a positive integer. If $a_n \leq b_n$ for all $n \geq N$, then we have*

$$\liminf_{n\to\infty} a_n \leq \liminf_{n\to\infty} b_n \quad and \quad \limsup_{n\to\infty} a_n \leq \limsup_{n\to\infty} b_n.$$

▌ 5.1.4 Cauchy Sequences and Complete Metric Spaces

Definition 5.12. *A sequence $\{p_n\}$ in a metric space X with metric d is called a **Cauchy sequence** if for every $\epsilon > 0$, there exists a positive integer N such that*

$$d(p_m, p_n) < \epsilon$$

whenever $m, n \geq N$.

The relationship between convergent sequences and Cauchy sequences in X can be characterized in the following theorem:

Theorem 5.13. *Let X be a metric space and $\{p_n\}$ be a sequence of X.*

(a) *If $\{p_n\}$ converges, then it is Cauchy.*

(b) *Suppose that X is compact or $\mathbb{R}^n$. If $\{p_n\}$ is Cauchy, then it is convergent.*

In other words, Theorem 5.13(a) says that the condition of a convergent sequence is stronger than that of a Cauchy sequence in *any* metric space. This kind of difference is due to the nature of the space the sequence lies and Theorem 5.13(b) indicates that they are equivalent if X is compact or the Euclidean space $\mathbb{R}^n$.

We remark that the converse of Theorem 5.13(a) is not true, see Problem 5.19 for an example. Thus, inspired by Theorem 5.13(b), it is natural to search more spaces where Cauchy sequences and convergent sequences are equivalent. In fact, such a metric space is called a **complete metric space**. For instances, every compact metric space and $\mathbb{R}^n$ is complete.

5.2 Convergence of Sequences

Problem 5.1

$(\star)$ *Evaluate the limit*

$$\lim_{n\to\infty} \left(\frac{1}{2} + \frac{3}{2^2} + \cdots + \frac{2n-1}{2^n} \right).$$

Proof. Suppose that

$$p_n = \frac{1}{2} + \frac{3}{2^2} + \cdots + \frac{2n-1}{2^n}. \tag{5.3}$$

Then we have

$$\frac{1}{2}p_n = \frac{1}{2^2} + \frac{3}{2^3} + \cdots + \frac{2n-1}{2^{n+1}}. \tag{5.4}$$

The subtraction of the expressions (5.3) and (5.4) gives

$$\frac{1}{2}p_n = \frac{1}{2} + \left(\frac{1}{2} + \frac{1}{2^2} + \cdots + \frac{1}{2^{n-1}} \right) - \frac{2n-1}{2^{n+1}} = \frac{1}{2} + 1 - \frac{1}{2^{n-1}} - \frac{2n-1}{2^{n+1}}$$

and then

$$p_n = 3 - \frac{1}{2^{n-2}} - \frac{2n-1}{2^n}$$

Hence, it follows from Theorem 5.5 and then Theorem 5.7(a) that

$$\lim_{n\to\infty} p_n = 3.$$

This completes the proof of the problem. ∎

Problem 5.2

$(\star)$ *Evaluate the limit*

$$\lim_{n\to\infty} \left[\frac{1}{1\cdot 2} + \frac{1}{2\cdot 3} + \cdots + \frac{1}{n(n+1)} \right].$$

Proof. Since

$$\frac{1}{1\cdot 2} + \frac{1}{2\cdot 3} + \cdots + \frac{1}{n(n+1)} = \left(1 - \frac{1}{2} \right) + \left(\frac{1}{2} - \frac{1}{3} \right) + \cdots + \left(\frac{1}{n} - \frac{1}{n+1} \right)$$

$$= 1 - \frac{1}{n+1},$$

we obtain from Theorem 5.5 and then Theorem 5.7(a) that

$$\lim_{n\to\infty} \left[\frac{1}{1\cdot 2} + \frac{1}{2\cdot 3} + \cdots + \frac{1}{n(n+1)} \right] = 1,$$

completing the proof of the problem. ∎

Problem 5.3

$(\star)$ $(\star)$ *Given that $\theta > 1$ and $k \in \mathbb{N}$. Prove that*

$$\lim_{n \to \infty} \frac{n^k}{\theta^n} = 0.$$

Proof. Since $\theta > 1$, there exists a $\delta > 0$ such that $\theta = 1 + \delta$. For a *fixed* k, if $n > k$, then the Binomial Theorem implies that

$$\theta^n = (1 + \delta)^n > \frac{n(n-1)\cdots(n-k)}{(k+1)!}\delta^{k+1}$$

so that

$$0 < \frac{n^k}{\theta^n} < \frac{n^k(k+1)!}{n(n-1)\cdots(n-k)\delta^{k+1}} = \frac{(k+1)!}{n(1-\frac{1}{n})(1-\frac{2}{n})\cdots(1-\frac{k}{n})\delta^{k+1}}. \tag{5.5}$$

By Theorem 5.7(a), we have $\lim\limits_{n \to \infty} \left(1 - \dfrac{j}{n}\right) = 1$ for $1 \le j \le k$, we see that

$$\lim_{n \to \infty} \frac{(k+1)!}{n(1-\frac{1}{n})(1-\frac{2}{n})\cdots(1-\frac{k}{n})\delta^{k+1}} = 0.$$

Finally, we apply Theorem 5.6 (Squeeze Theorem for Convergent Sequences) to the inequalities (5.5) to get

$$\lim_{n \to \infty} \frac{n^k}{\theta^n} = 0$$

which is our desired result. This finishes the proof of the problem. ∎

Problem 5.4

$(\star)$ *Given that $a_k \ge 0$ for $k = 1, 2, \ldots, m$. Evaluate the limit*

$$\lim_{n \to \infty} \sqrt[n]{a_1^n + \cdots + a_m^n}.$$

Proof. By rearrangement if necessary, we may assume that $a_1 \le a_2 \le \cdots \le a_m$. Thus we have

$$a_m^n \le a_1^n + \cdots + a_m^n \le ma_m^n$$

and then

$$a_m \le \sqrt[n]{a_1^n + \cdots + a_m^n} \le m^{\frac{1}{n}}a_m. \tag{5.6}$$

By Theorem 5.7(a), we have $\lim\limits_{n \to \infty} m^{\frac{1}{n}} = 1$. Apply Theorem 5.6 (Squeeze Theorem for Convergent Sequences) to the inequalities (5.6), we establish that

$$\lim_{n \to \infty} \sqrt[n]{a_1^n + \cdots + a_m^n} = a_m,$$

completing the proof of the problem. ∎

Problem 5.5

$(\star)$ Given that $\ln(1 + \frac{1}{n}) < \frac{1}{n}$ for every $n \in \mathbb{N}$. Suppose that

$$p_n = \left(1 + \frac{1}{3}\right)\left(1 + \frac{1}{9}\right)\cdots\left(1 + \frac{1}{3^n}\right), \tag{5.7}$$

where $n = 1, 2, \ldots$. Prove that $\{p_n\}$ converges.

Proof. Since each $1 + \frac{1}{3^n} > 0$ for all $n \in \mathbb{N}$, we see easily that $\{p_n\}$ is monotonic increasing. Furthermore, we can take the logarithm to both sides of (5.7) to get

$$\ln p_n = \ln\left(1 + \frac{1}{3}\right) + \ln\left(1 + \frac{1}{3^2}\right) + \cdots + \ln\left(1 + \frac{1}{3^n}\right). \tag{5.8}$$

By the given hint, we get the following inequality from the expression (5.8) that

$$\ln p_n < \frac{1}{3} + \frac{1}{3^2} + \cdots + \frac{1}{3^n} < 1 + \frac{1}{3} + \cdots = \frac{3}{2}$$

which means that $\{p_n\}$ is bounded above by $\frac{3}{2}$. By the Monotonic Convergence Theorem, we know that $\{p_n\}$ is convergent and this completes the proof of the problem. ∎

Problem 5.6

$(\star)$ Define $p_1 = \sqrt{3}$ and for $n = 2, 3, \ldots$, we have

$$p_n = \underbrace{\sqrt{3 + \sqrt{3 + \cdots + \sqrt{3}}}}_{n \text{ square root symbols}}.$$

Prove that $\{p_n\}$ converges.

Proof. It can be shown by induction that $\{p_n\}$ is monotonically increasing and bounded above by 3. By the Monotonic Convergence Theorem, we see that $\{p_n\}$ is convergent. We finish the proof of the problem. ∎

Problem 5.7

$(\star)(\star)$ Suppose that $\lim\limits_{n \to \infty} \cos \delta_n = 1$ if $\lim\limits_{n \to \infty} \delta_n = 0$. For every positive integer n, we define

$$p_n = \cos(2n!e\pi).$$

Prove that

$$\lim_{n \to \infty} p_n = 2\pi.$$

Proof. By definition, we have

$$e = \left(1 + \frac{1}{1!} + \cdots + \frac{1}{n!}\right) + \left[\frac{1}{(n+1)!} + \cdots\right].$$

so that

$$2n!e = 2n!\left(1 + \frac{1}{1!} + \cdots + \frac{1}{n!}\right) + 2\left[\frac{1}{n+1} + \frac{1}{(n+1)(n+2)} + \cdots\right] = 2N_n + \delta_n, \qquad (5.9)$$

where $N_n = n!\left(1 + \frac{1}{1!} + \cdots + \frac{1}{n!}\right)$ and $\delta_n = 2\left[\frac{1}{n+1} + \frac{1}{(n+1)(n+2)} + \cdots\right]$. Since

$$\frac{2}{n+1} < \delta_n < 2\left[\frac{1}{n+1} + \frac{1}{(n+1)^2} + \cdots\right] = \frac{2}{n},$$

Theorem 5.6 (Squeeze Theorem for Convergent Sequences) shows that

$$\lim_{n\to\infty} \delta_n = 0.$$

Note that N_n is always a positive integer for every $n \in \mathbb{N}$, so we follow from the expression (5.9) and the fact $\cos(2n\pi + \theta) = \cos\theta$ that

$$\cos(2n!e\pi) = \cos(2N_n\pi + \delta_n\pi) = \cos\delta_n\pi.$$

Hence we conclude from the given hint that

$$\lim_{n\to\infty} \cos(2n!e\pi) = \lim_{n\to\infty} \cos\delta_n\pi = 1,$$

completing the proof of the problem.   ■

Problem 5.8

(⋆) *Suppose that $\{p_n\}$ is a sequence of nonnegative real numbers with $p_2 = 0$. If $p_{m+n} \leq p_m + p_n$ for all $m, n \in \mathbb{N}$, prove that*

$$\lim_{n\to\infty} \frac{p_n}{n} = 0 \qquad (5.10)$$

Proof. Take $m = 2$. For any large positive integer n, we have $n = 2q + r$, where $q \in \mathbb{N}$ and $0 \leq r \leq 1$. By the hypothesis, we see that

$$0 \leq p_n = p_{2q+r} \leq p_{2+2+\cdots+2} + p_r \leq qp_2 + p_r \leq p_r$$

which implies that

$$0 \leq \frac{p_n}{n} \leq \frac{p_r}{n}. \qquad (5.11)$$

Since $\lim_{n\to\infty} \frac{p_r}{n} = 0$, we apply Theorem 5.6 (Squeeze Theorem for Convergent Sequences) to the inequalities (5.11) to obtain the required result (5.10). This ends the proof of the problem.   ■

Problem 5.9

$(\star)$ Given that $\{p_n\}$ is a sequence of real numbers. If $p_n \to p$ as $n \to \infty$, prove that $|p_n| \to |p|$ as $n \to \infty$.

Proof. Given $\epsilon > 0$. Since $p_n \to p$, there exists a positive integer N such that $|p_n - p| < \epsilon$ for all $n \geq N$. With this same N, it deduces from Problem 2.5 that

$$||p_n| - |p|| \leq |p_n - p| < \epsilon$$

for all $n \geq N$. Hence we have completed the proof of the problem. ∎

Problem 5.10

$(\star)(\star)$ Suppose that $\{A_n\}$ and $\{B_n\}$ are sequences of real numbers converging to A and B respectively. Let

$$C_n = \frac{A_0 B_n + A_1 B_{n-1} + \cdots + A_n B_0}{n+1}.$$

Prove that $\lim\limits_{n \to \infty} C_n = AB$.

Proof. By Theorem 5.2, $\{A_n\}$ and $\{B_n\}$ are bounded. Let M be a positive number such that

$$|A_n| \leq M \quad \text{and} \quad |B_n| \leq M \tag{5.12}$$

for all $n \in \mathbb{N}$. Since $A_n \to A$ and $B_n \to B$, for every $\epsilon > 0$, we know from the definition that there exists a positive integer N such that

$$|A_n - A| < \frac{\epsilon}{2(M + |A|)} \quad \text{and} \quad |B_n - B| < \frac{\epsilon}{2(M + |A|)} \tag{5.13}$$

for all $n \geq N$. Now we deduce from the inequalities (5.12) that

$$\begin{aligned}
|A_k B_{n-k} - AB| &= |A_k B_{n-k} - AB_{n-k} + AB_{n-k} - AB| \\
&\leq |A_k - A||B_{n-k}| + |A||B_{n-k} - B| \\
&\leq M|A_k - A| + |A||B_{n-k} - B|. \tag{5.14}
\end{aligned}$$

for all $k \in \mathbb{N}$.

Consider the positive integers n and k with $n \geq N^2$ and $[\sqrt{n}] \leq k \leq n - [\sqrt{n}]$.[e] By these, we know that

$$k \geq [\sqrt{n}] \geq N \quad \text{and} \quad n - k \geq \sqrt{n} \geq N.$$

Thus we follow from the inequalities (5.13) and (5.14) that

$$|A_k B_{n-k} - AB| < \frac{\epsilon}{2} \tag{5.15}$$

for $n \geq N^2$ and $[\sqrt{n}] \leq k \leq n - [\sqrt{n}]$.

---

[e]Recall that $[x]$ denotes the greatest integer less than or equal to x.

For $0 \leq k \leq \sqrt{n}$, we have

$$\sum_{k=0}^{[\sqrt{n}]} |A_k B_{n-k} - AB| \leq \sum_{k=0}^{[\sqrt{n}]} (M^2 + |AB|) \leq (M^2 + |AB|)([\sqrt{n}] + 1) \tag{5.16}$$

and similarly, for $n - [\sqrt{n}] \leq k \leq n$, we have

$$\sum_{k=n-[\sqrt{n}]}^{n} |A_k B_{n-k} - AB| \leq \sum_{k=n-[\sqrt{n}]}^{n} (M^2 + |AB|) \leq (M^2 + |AB|)([\sqrt{n}] + 1). \tag{5.17}$$

Finally, by combining the inequalities (5.15), (5.16) and (5.17), we derive that

$$\begin{aligned}
|C_n - AB| &\leq \left| \frac{1}{n+1} \sum_{k=0}^{n} A_k B_{n-k} - AB \right| \\
&\leq \frac{1}{n+1} \left| \sum_{k=0}^{n} A_k B_{n-k} - (n+1)AB \right| \\
&= \frac{1}{n+1} \left| \sum_{k=0}^{n} (A_k B_{n-k} - AB) \right| \\
&\leq \frac{1}{n+1} \sum_{k=0}^{[\sqrt{n}]} |A_k B_{n-k} - AB| + \frac{1}{n+1} \sum_{k=[\sqrt{n}]}^{n-[\sqrt{n}]} |A_k B_{n-k} - AB| \\
&\quad + \frac{1}{n+1} \sum_{k=n-[\sqrt{n}]}^{n} |A_k B_{n-k} - AB| \\
&\leq \frac{2(M^2 + |AB|)([\sqrt{n}] + 1)}{n+1} + \frac{\epsilon}{2}.
\end{aligned} \tag{5.18}$$

Hence, if we take n large enough[f], then we can make

$$\frac{2(M^2 + |AB|)([\sqrt{n}] + 1)}{n+1} < \frac{\epsilon}{2}.$$

By this and the inequality (5.18), we conclude that

$$|C_n - AB| < \epsilon$$

for large enough n and this proves the desired result. This completes the proof of the problem. ∎

Problem 5.11

⋆ ⋆ *Suppose that $\{p_n\}$ is a sequence of real numbers such that $p_n - p_{n-2} \to 0$. Prove that*

$$\lim_{n \to \infty} \frac{p_n - p_{n-1}}{n} = 0. \tag{5.19}$$

---

[f]Recall that once we choose ϵ, the positive integer N is fixed and then we can take any large integer n satisfying $n \geq N$.

Proof. Given $\epsilon > 0$, since $p_n - p_{n-2} \to 0$, there exists a positive integer N such that

$$|p_n - p_{n-2}| < \frac{\epsilon}{2} \tag{5.20}$$

for $n \geq N$. We fix this N. It is clear that

$$\begin{aligned}
p_n - p_{n-1} &= (p_n - p_{n-2} + p_{n-2} - p_{n-1}) \\
&= (p_n - p_{n-2}) - (p_{n-1} - p_{n-3} + p_{n-3} - p_{n-2}) \\
&= (p_n - p_{n-2}) - (p_{n-1} - p_{n-3}) - (p_{n-3} - p_{n-4} + p_{n-4} - p_{n-2}) \\
&= (p_n - p_{n-2}) - (p_{n-1} - p_{n-3}) + (p_{n-2} - p_{n-4}) - (p_{n-3} - p_{n-4}) \\
&= (p_n - p_{n-2}) - (p_{n-1} - p_{n-3}) + (p_{n-2} - p_{n-4}) - \cdots \\
&\quad + (-1)^N (p_N - p_{N-2} + p_{N-2} - p_{N-1}). \tag{5.21}
\end{aligned}$$

Thus we deduce from the inequalities (5.20) and (5.21) that

$$\begin{aligned}
|p_n - p_{n-1}| &\leq |p_n - p_{n-2}| + |p_{n-1} - p_{n-3}| + |p_{n-2} - p_{n-4}| + \cdots \\
&\quad + |p_N - p_{N-2}| + |p_{N-2} - p_{N-1}| \\
&< (n - N + 1)\epsilon + |p_{N-2} - p_{N-1}|
\end{aligned}$$

for $n \geq N$ and thus

$$\left| \frac{p_n - p_{n-2}}{n} \right| < \frac{(n - N + 1)\epsilon}{2n} + \frac{|p_{N-2} - p_{N-1}|}{n} < \frac{\epsilon}{2} + \frac{\epsilon}{2} = \epsilon$$

for large enough n because N is *fixed*. By definition, we have the limit (5.19) which completes the proof of the problem. ∎

5.3 Upper and Lower Limits

Problem 5.12

(⋆) *Suppose that $a_n = (-1)^n (2 + \frac{3}{n})$ for $n = 1, 2, \ldots$. Find*

$$\limsup_{n \to \infty} a_n \quad \text{and} \quad \liminf_{n \to \infty} a_n.$$

Proof. It is clear that

$$a_{2n} = 2 + \frac{3}{2n} \quad \text{and} \quad a_{2n-1} = -2 - \frac{3}{2n - 1}$$

for $n = 1, 2, \ldots$. Therefore, we have

$$\limsup_{n \to \infty} a_n = \lim_{n \to \infty} a_{2n} = 2 \quad \text{and} \quad \liminf_{n \to \infty} a_n = \lim_{n \to \infty} a_{2n-1} = -2.$$

Hence we complete the proof of the problem. ∎

Problem 5.13

$\star$ $\star$ *If $\{a_n\}$ is a sequence of positive real numbers, prove that*

$$\liminf_{n \to \infty} \frac{a_{n+1}}{a_n} \le \liminf_{n \to \infty} \sqrt[n]{a_n} \le \limsup_{n \to \infty} \sqrt[n]{a_n} \le \limsup_{n \to \infty} \frac{a_{n+1}}{a_n}.$$

Proof. We only prove the case

$$\limsup_{n \to \infty} \sqrt[n]{a_n} \le \limsup_{n \to \infty} \frac{a_{n+1}}{a_n}$$

because the case $\liminf\limits_{n \to \infty} \dfrac{a_{n+1}}{a_n} \le \liminf\limits_{n \to \infty} \sqrt[n]{a_n}$ is very similar. Let $A = \limsup\limits_{n \to \infty} \dfrac{a_{n+1}}{a_n}$. If $A = \infty$, then there is nothing to prove.

Suppose that $A < \infty$. Given $\epsilon > 0$. By Theorem 5.9(b), there exists a positive integer N such that

$$\frac{a_{n+1}}{a_n} < A + \epsilon \tag{5.22}$$

for all $n \ge N$. For $n > N$, we deduce from the inequality (5.22) that

$$0 < a_n = \frac{a_n}{a_{n-1}} \times \frac{a_{n-1}}{a_{n-2}} \times \cdots \times \frac{a_{N+1}}{a_N} \times a_N < (A+\epsilon)^{n-N} a_N = (A+\epsilon)^n \times \frac{a_N}{(A+\epsilon)^N}$$

so that

$$0 < \sqrt[n]{a_n} < (A+\epsilon) \times \sqrt[n]{\frac{a_N}{(A+\epsilon)^N}}. \tag{5.23}$$

Since $\frac{a_N}{(A+\epsilon)^N} > 0$, it follows from Theorem 5.7(a) that

$$\lim_{n \to \infty} \sqrt[n]{\frac{a_N}{(A+\epsilon)^N}} = 1$$

and we deduce from the inequalities (5.23) and Theorem 5.11 that

$$\limsup_{n \to \infty} \sqrt[n]{a_n} \le (A+\epsilon) \limsup_{n \to \infty} \sqrt[n]{\frac{a_N}{(A+\epsilon)^N}} = (A+\epsilon) \lim_{n \to \infty} \sqrt[n]{\frac{a_N}{(A+\epsilon)^N}} = A + \epsilon.$$

Since ϵ is arbitrary, we must have

$$\limsup_{n \to \infty} \sqrt[n]{a_n} \le A = \limsup_{n \to \infty} \frac{a_{n+1}}{a_n},$$

completing the proof of the problem.   ∎

Problem 5.14

$\star$ $\star$ *Prove that*

$$\lim_{n \to \infty} \frac{\sqrt[n]{n!}}{n} = \frac{1}{e}.$$

Proof. Put $a_n = \frac{n!}{n^n}$ into Problem 5.13, we have

$$\frac{a_{n+1}}{a_n} = \frac{(n+1)!}{(n+1)^{n+1}} \times \frac{n^n}{n!} = \left(\frac{n}{n+1}\right)^n \quad \text{and} \quad \sqrt[n]{a_n} = \frac{\sqrt[n]{n!}}{n}.$$

Recall the definition[g]

$$e^x = \lim_{n \to \infty} \left(1 + \frac{x}{n}\right)^n$$

for every $x \in \mathbb{R}$. By this, we have

$$\limsup_{n \to \infty} \frac{a_{n+1}}{a_n} = \liminf_{n \to \infty} \frac{a_{n+1}}{a_n} = \lim_{n \to \infty} \left(\frac{n}{n+1}\right)^n = \lim_{n \to \infty} \left(1 - \frac{1}{n+1}\right)^n = e^{-1}. \qquad (5.24)$$

Thus it follows from the result (5.24) and Theorem 5.6 (Squeeze Theorem for Convergent Sequences) that

$$\limsup_{n \to \infty} \sqrt[n]{a_n} = \liminf_{n \to \infty} \sqrt[n]{a_n} = \frac{1}{e}.$$

Hence we obtain from Theorem 5.10 that

$$\lim_{n \to \infty} \frac{\sqrt[n]{n!}}{n} = \lim_{n \to \infty} \sqrt[n]{a_n} = \frac{1}{e},$$

completing the proof of the problem. ∎

Problem 5.15

$\bigstar$ $\bigstar$ *Suppose that $\{x_n\}$ is a convergent real sequence. Let*

$$a_n = \frac{x_1 + \cdots + x_n}{n}$$

for $n = 1, 2, \ldots$. Prove that

$$\lim_{n \to \infty} a_n = \lim_{n \to \infty} x_n. \qquad (5.25)$$

Proof. Suppose that $\lim_{n \to \infty} x_n = x$. Let $x_n = x + \xi_n$, where $\xi_n \to 0$ as $n \to \infty$. It is easy to see that

$$a_n - x = \frac{x_1 + \cdots + x_n}{n} - x$$

$$= \frac{1}{n}[(x_1 - x) + (x_2 - x) + \cdots + (x_n - x)]$$

$$= \frac{1}{n}(\xi_1 + \xi_2 + \cdots + \xi_n). \qquad (5.26)$$

Since $\xi_n \to 0$ as $n \to \infty$, given $\epsilon > 0$, there exists a positive integer N such that

$$-\epsilon < \xi_n < \epsilon$$

for all $n > N$. We *fix* this N, so we deduce from the inequality (5.26) that

$$\frac{1}{n}(\xi_1 + \cdots + \xi_N) - \epsilon < a_n - x < \frac{1}{n}(\xi_1 + \cdots + \xi_N) + \epsilon$$

---

[g]See, for example, [7, §215, Eqn. (1)].

for all $n > N$. By Theorem 5.11, we have

$$-\epsilon \le \liminf_{n\to\infty}(a_n - x) \le \limsup_{n\to\infty}(a_n - x) \le \epsilon. \tag{5.27}$$

Since ϵ is arbitrary, the inequality (5.27) means that

$$\liminf_{n\to\infty}(a_n - x) = \limsup_{n\to\infty}(a_n - x) = 0$$

which proves the desire inequality (5.25) by Theorem 5.10. This ends the proof of the problem.

∎

Problem 5.16

⋆ ⋆ *Let $\{a_n\}$ be a sequence of positive numbers. If*

$$\limsup_{n\to\infty} a_n \times \limsup_{n\to\infty} \frac{1}{a_n} = 1, \tag{5.28}$$

prove that $\{\frac{1}{a_n}\}$ is convergent.

Proof. We know that the left-hand side of the hypothesis (5.28) cannot be $0\cdot\infty$, so the limits must be nonzero real numbers. Let $\alpha = \limsup\limits_{n\to\infty} a_n$. By Theorem 5.9(a), *there exists* a subsequence $\{a_{n_k}\}$ such that

$$a_{n_k} \to \alpha \tag{5.29}$$

as $k \to \infty$. Let

$$\liminf_{n\to\infty} \frac{1}{a_n} = \beta.$$

By the definition (5.2), we have

$$\beta \le \lim_{j\to\infty} \frac{1}{a_{n_j}}$$

for all subsequences $\{a_{n_j}\}$. In particular, we know from the limit (5.29) that $\beta \le \frac{1}{\alpha}$. Assume that $\beta < \frac{1}{\alpha}$. Then there exists a $\delta > 0$ such that

$$\beta < \beta + \delta < \frac{1}{\alpha}.$$

Since $\beta + \delta$ is *not* a lower bound of the set of all subsequential limits of $\{\frac{1}{a_n}\}$ anymore, we can find a subsequence $\{\frac{1}{a_{n_j}}\}$ such that

$$\lim_{j\to\infty} \frac{1}{a_{n_j}} < \beta + \delta < \frac{1}{\alpha}$$

which implies that

$$\alpha < \lim_{j\to\infty} a_{n_j}$$

but this contradicts the definition that α is the least upper bound of the set of all subsequential limits of $\{a_n\}$. Therefore, we must have $\beta = \frac{1}{\alpha}$, i.e.,

$$\limsup_{n\to\infty} a_n \times \liminf_{n\to\infty} \frac{1}{a_n} = 1. \tag{5.30}$$

By comparing the results (5.28) and (5.30), we get

$$\limsup_{n \to \infty} \frac{1}{a_n} = \liminf_{n \to \infty} \frac{1}{a_n} = \frac{1}{\alpha}.$$

Hence, by Theorem 5.10, we have proven that $\{\frac{1}{a_n}\}$ converges and this completes the proof of the problem. ∎

Problem 5.17

⋆ ⋆ ⋆ *Suppose that $a_n \geq 0$ and $b_n \geq 0$ for all $n = 1, 2, \ldots$. Show that*

$$\liminf_{n \to \infty} a_n \times \liminf_{n \to \infty} b_n \leq \liminf_{n \to \infty}(a_n \times b_n) \leq \liminf_{n \to \infty} a_n \times \limsup_{n \to \infty} b_n.$$

Proof. Let $A = \liminf\limits_{n \to \infty} a_n$. We prove the inequality

$$\liminf_{n \to \infty}(a_n \times b_n) \leq \liminf_{n \to \infty} a_n \times \limsup_{n \to \infty} b_n \tag{5.31}$$

first. By the definition (5.2), there exists a subsequence $\{a_{n_k}\}$ such that

$$\lim_{k \to \infty} a_{n_k} = A. \tag{5.32}$$

For the corresponding subsequence $\{b_{n_k}\}$, let $B = \limsup\limits_{k \to \infty} b_{n_k}$. Then there exists a subsequence $\{b_{n_{k_j}}\}$ (of $\{b_{n_k}\}$) such that

$$\lim_{j \to \infty} b_{n_{k_j}} = B. \tag{5.33}$$

It is clear that

$$B = \limsup_{k \to \infty} b_{n_k} \leq \limsup_{n \to \infty} b_n. \tag{5.34}$$

Apply Theorem 5.10 to the limit (5.32), we know that

$$\lim_{j \to \infty} a_{n_{k_j}} = A, \tag{5.35}$$

where $\{a_{n_{k_j}}\}$ is *any* subsequence of $\{a_{n_k}\}$. Combining the limits (5.33) and (5.35), we achieve

$$\lim_{j \to \infty}(a_{n_{k_j}} \times b_{n_{k_j}}) = AB.$$

By the definition (5.2) again, we have

$$\liminf_{n \to \infty}(a_n \times b_n) \leq AB. \tag{5.36}$$

Since $a_n \geq 0$ and $b_n \geq 0$ for all $n = 1, 2, \ldots$, A and B must be nonnegative. Substitute the inequality (5.34) into the inequality (5.36), we are able to obtain the desired inequality (5.31).

Next, we prove the inequality

$$\liminf_{n \to \infty} a_n \times \liminf_{n \to \infty} b_n \leq \liminf_{n \to \infty}(a_n \times b_n). \tag{5.37}$$

Let $C = \liminf\limits_{n \to \infty} (a_n \times b_n)$. If $A = 0$, then there is nothing to prove. Thus we may assume that $A > 0$. By the analogous result of Theorem 5.9, there exists a positive integer N such that

$$a_n > 0 \tag{5.38}$$

for all $n \geq N$. Again the analogous result of Theorem 5.9 implies that a subsequence $\{a_{n_k} b_{n_k}\}$ of $\{a_n b_n\}$ exists such that

$$\lim_{k \to \infty} (a_{n_k} b_{n_k}) = C. \tag{5.39}$$

For the sequence $\{a_{n_k}\}$, we can find a subsequence $\{a_{n_{k_j}}\}$ such that

$$\lim_{j \to \infty} a_{n_{k_j}} = A' = \liminf_{k \to \infty} a_{n_k}. \tag{5.40}$$

By the definition of A, A' and the lower limit, we have

$$0 < A = \liminf_{n \to \infty} a_n \leq \liminf_{k \to \infty} a_{n_k} = A'. \tag{5.41}$$

Then it follows from the inequality (5.38), the limits (5.39) and (5.40) with the application of Theorem 5.10 to the limit (5.39) that

$$\lim_{j \to \infty} b_{n_{k_j}} = \lim_{j \to \infty} \left[(a_{n_{k_j}} b_{n_{k_j}}) \times \frac{1}{a_{n_{k_j}}} \right] = \frac{C}{A'}.$$

By the definition (5.2), this implies that

$$\liminf_{n \to \infty} b_n \leq \frac{C}{A'}. \tag{5.42}$$

Hence, after substituting the definition of A' and C back into the inequality (5.42) and then using the inequality (5.41), we obtain the expected inequality (5.37), completing the proof of the problem. ∎

Remark 5.1

Similar to Problem 5.17, if $a_n \geq 0$ and $b_n \geq 0$ for all $n = 1, 2, \ldots$, then we can prove that

$$\liminf_{n \to \infty} a_n \times \limsup_{n \to \infty} b_n \leq \limsup_{n \to \infty} (a_n \times b_n) \leq \limsup_{n \to \infty} a_n \times \limsup_{n \to \infty} b_n.$$

Problem 5.18

⋆ ⋆ *Reprove Problem 5.16 by using Problem 5.17 and Remark 5.1.*

Proof. Since $a_n > 0$ for all $n = 1, 2, \ldots$, we have $\frac{1}{a_n} > 0$ for all $n = 1, 2, \ldots$. In other words, the sequences $\{a_n\}$ and $\{\frac{1}{a_n}\}$ satisfy the hypotheses of Problem 5.17 and Remark 5.1. Put $b_n = \frac{1}{a_n}$ into Problem 5.17 and Remark 5.1, we have

$$\liminf_{n \to \infty} a_n \times \liminf_{n \to \infty} \frac{1}{a_n} \leq 1 \leq \liminf_{n \to \infty} a_n \times \limsup_{n \to \infty} \frac{1}{a_n} \tag{5.43}$$

and

$$\liminf_{n \to \infty} a_n \times \limsup_{n \to \infty} \frac{1}{a_n} \le 1 \le \limsup_{n \to \infty} a_n \times \limsup_{n \to \infty} \frac{1}{a_n} \quad (5.44)$$

respectively. Combining the inequalities (5.43) and (5.44), it is easy to see that

$$1 \le \liminf_{n \to \infty} a_n \times \limsup_{n \to \infty} \frac{1}{a_n} \le 1$$

which means that

$$\liminf_{n \to \infty} a_n \times \limsup_{n \to \infty} \frac{1}{a_n} = 1.$$

By this and the hypothesis (5.28), we have $\limsup_{n \to \infty} \frac{1}{a_n} \ne 0$ and then we derive that

$$\liminf_{n \to \infty} a_n = \limsup_{n \to \infty} a_n.$$

Hence the sequence $\{a_n\}$ and then the sequence $\{\frac{1}{a_n}\}$ are convergent and we finish the proof of the problem. ∎

5.4 Cauchy Sequences and Complete Metric Spaces

Problem 5.19

(⋆) Suppose that $X = \mathbb{R}^+$, the set of all positive real numbers, and the metric d of X is given by

$$d(x, y) = |x - y| \quad (5.45)$$

for all $x, y \in X$. Prove that if $p_n = \frac{1}{n}$ for all $n = 1, 2, \ldots$, then $\{p_n\}$ is Cauchy but not convergent in X.

Proof. Given that $\epsilon > 0$. By Theorem 2.1 (The Archimedean Property), there exists a positive integer N such that

$$\frac{1}{N} < \frac{\epsilon}{2}. \quad (5.46)$$

It is clear that the inequality (5.46) holds for every $n \ge N$. Therefore, if $m, n \ge N$, then we have

$$d(p_n, p_m) = \left| \frac{1}{n} - \frac{1}{m} \right| \le \frac{1}{n} + \frac{1}{m} < \frac{\epsilon}{2} + \frac{\epsilon}{2} = \epsilon.$$

By Definition 5.12, the sequence $\{p_n\}$ is Cauchy. However, since its limit point is $0 \notin X$, it does not satisfy the definition of a convergent sequence (see §5.1.1). Hence we have completed the proof of the problem. ∎

Remark 5.2

In other words, Problem 5.19 shows that the space $\mathbb{R}^+$ with the metric d given by (5.45) is *not* complete.

Problem 5.20

$(\star)$ *Suppose that $\{a_n\}$ is a bounded real sequence and $|x| < 1$. We define $\{p_n\}$ by*

$$p_n = a_n x^n + \cdots + a_1 x + a_0$$

for $n = 1, 2, \ldots$. Prove that $\{p_n\}$ is Cauchy.

Proof. Since $\{a_n\}$ is bounded, there exists a positive constant M such that

$$|a_n| \leq M$$

for all $n = 1, 2, \ldots$. Given $\epsilon > 0$. Let N be a positive integer such that

$$N > \frac{1}{\log |x|} \times \log \frac{\epsilon(1 - |x|)}{M} - 1. \tag{5.47}$$

Then for all $m > n \geq N$, we follow from the inequality (5.47) that

$$
\begin{aligned}
|p_m - p_n| &= |a_m x^m + \cdots + a_{n+1} x^{n+1}| \\
&\leq M(|x|^m + |x|^{m-1} + \cdots + |x|^{n+1}) \\
&\leq M|x|^{n+1}(1 + |x| + \cdots + |x|^{m-n-1}) \\
&< M|x|^{n+1}(1 + |x| + \cdots) \\
&= \frac{M|x|^{n+1}}{1 - |x|} \\
&\leq \frac{M|x|^{N+1}}{1 - |x|} \\
&< \epsilon.
\end{aligned}
$$

Hence $\{p_n\}$ is Cauchy by Definition 5.12 and we finish the proof of the problem. ∎

Problem 5.21

$(\star)$ *Suppose that $\{p_n\}$ is a real sequence defined by*

$$p_n = \frac{n+1}{n-2}$$

for $n = 3, 4, \ldots$. Prove that $\{p_n\}$ is Cauchy.

Proof. We note that

$$
\begin{aligned}
|p_m - p_n| &= \left| \frac{m+1}{m-2} - \frac{n+1}{n-2} \right| \\
&= \left| \frac{3(n-m)}{(m-2)(n-2)} \right| \\
&\leq \left| \frac{3n}{(m-2)(n-2)} \right| + \left| \frac{3m}{(m-2)(n-2)} \right|. \tag{5.48}
\end{aligned}
$$

Since $\frac{3n}{n-2}$ and $\frac{3m}{m-2}$ are obviously bounded, we can find a positive constant M such that

$$\left|\frac{3n}{n-2}\right| \le M \quad \text{and} \quad \left|\frac{3m}{m-2}\right| \le M.$$

Put these into the inequality (5.48), we achieve

$$|p_m - p_n| \le M\left(\frac{1}{m-2} + \frac{1}{n-2}\right). \tag{5.49}$$

Therefore, if we let N to be a positive integer such that $N > \frac{2M}{\epsilon} + 2$, then for $m, n \ge N$, we derive from the inequality (5.49) that

$$|p_m - p_n| \le M\left(\frac{1}{m-2} + \frac{1}{n-2}\right) < \frac{\epsilon}{2} + \frac{\epsilon}{2} = \epsilon.$$

Hence the sequence $\{p_n\}$ is Cauchy and we finish the proof of the problem. ∎

Problem 5.22

(⋆) Suppose that $\{p_n\}$ is a real sequence defined by

$$p_n = \frac{\sin(1!)}{1 \times 2} + \frac{\sin(2!)}{2 \times 3} + \cdots + \frac{\sin(n!)}{n \times (n+1)}$$

for $n = 1, 2, \ldots$. Determine the convergence of $\{p_n\}$.

Proof. Without the knowledge of the limit (if it exists) of $\{p_n\}$, we prove that it is Cauchy in $\mathbb{R}$ so that it converges in $\mathbb{R}$ by Theorem 5.13(b). To this end, given $\epsilon > 0$. Let N be a positive integer such that

$$N > \frac{1}{\epsilon} - 1. \tag{5.50}$$

Then for $m > n \ge N$, we obtain from the inequality (5.50) and the fact $|\sin x| \le 1$ that

$$
\begin{aligned}
|p_m - p_n| &= \left|\frac{\sin(m!)}{m(m+1)} + \frac{\sin[(m-1)!]}{(m-1)m} + \cdots + \frac{\sin[(n+1)!]}{(n+1)(n+2)}\right| \\
&\le \frac{1}{m(m+1)} + \frac{1}{(m-1)m} + \cdots + \frac{1}{(n+1)(n+2)} \\
&= \left(\frac{1}{m} - \frac{1}{m+1}\right) + \left(\frac{1}{m-1} - \frac{1}{m}\right) + \cdots + \left(\frac{1}{n+1} - \frac{1}{n+2}\right) \\
&= \frac{1}{n+1} - \frac{1}{m+1} \\
&< \frac{1}{n+1} \\
&< \epsilon
\end{aligned}
$$

By Definition 5.12, $\{p_n\}$ is Cauchy and by Theorem 5.13(b), it is convergent in $\mathbb{R}$. This completes the proof of the problem. ∎

Problem 5.23

$(\star)$ *Suppose that $\{p_n\}$ is a real sequence defined by*

$$p_n = 1 + \frac{1}{2} + \cdots + \frac{1}{n}$$

for $n = 1, 2, \ldots$. Determine the convergence of $\{p_n\}$.

Proof. Take $\epsilon = \frac{1}{2}$. Consider

$$
\begin{aligned}
|p_{2n} - p_n| &= \frac{1}{n+1} + \frac{1}{n+2} + \cdots + \frac{1}{2n} \\
&> \underbrace{\frac{1}{2n} + \frac{1}{2n} + \cdots + \frac{1}{2n}}_{n \text{ terms}} \\
&= \frac{1}{2}.
\end{aligned}
$$

Thus there is *no* integer N such that $|p_m - p_n| < \frac{1}{2}$ for all $m, n \geq N$. By Definition 5.12, $\{p_n\}$ is not Cauchy and by Theorem 5.13, it is not convergent in $\mathbb{R}$. This ends the proof of the problem. ∎

Problem 5.24

$(\star)(\star)$ *Suppose that $\{p_n\}$ is Cauchy in a metric space X with metric d. If a subsequence $\{p_{n_k}\}$ converges, prove that $\{p_n\}$ is convergent in X.*

Proof. Suppose that $\lim\limits_{k \to \infty} p_{n_k} = p$. Given $\epsilon > 0$. Since $p_{n_k} \to p$, there exists a positive integer N_1 such that $k \geq N_1$ implies

$$d(p_{n_k}, p) < \frac{\epsilon}{2}. \tag{5.51}$$

Since $\{p_n\}$ is Cauchy, there is a positive integer N_2 such that $m, n \geq N_2$ implies that

$$d(p_m, p_n) < \frac{\epsilon}{2}. \tag{5.52}$$

Note that if $k \to \infty$, then $n_k \to \infty$. Therefore, k can be chosen large enough so that $n_k \geq N_2$. Then for $n \geq N_2$, we deduce from the inequalities (5.51) and (5.52) that

$$d(p_n, p) \leq d(p_n, p_{n_k}) + d(p_{n_k}, p) < \frac{\epsilon}{2} + \frac{\epsilon}{2} = \epsilon.$$

By definition, $\{p_n\}$ also converges. This ends the proof of the problem. ∎

Problem 5.25 (The Bolzano-Weierstrass Theorem)

$(\star)(\star)$ *Suppose that $\{x_n\}$ is a bounded infinite sequence of real numbers. Prove that there is a convergent subsequence $\{x_{n_k}\}$.*

Proof. Without loss of generality, we may assume that $\{x_n\} \subseteq [0, 1]$. Since $[0, 1] = [0, \frac{1}{2}] \cup [\frac{1}{2}, 1]$ and $\{x_n\}$ is infinite, at least one of $[0, \frac{1}{2}]$ and $[\frac{1}{2}, 1]$ contains infinitely many elements of $\{x_n\}$. Let this interval be I_1 and $\{x_{n^{(1)}}\}$ be the subsequence of $\{x_n\}$ such that

$$\{x_{n^{(1)}}\} \subseteq I_1$$

and the length of I_1 is $\frac{1}{2}$. Next, since

$$I_1 = \left[0, \frac{1}{4}\right] \cup \left[\frac{1}{4}, \frac{1}{2}\right] \quad \left(\text{or } I_1 = \left[\frac{1}{2}, \frac{3}{4}\right] \cup \left[\frac{3}{4}, 1\right]\right)$$

and $\{x_{n^{(1)}}\}$ is infinite, at least one of $[0, \frac{1}{4}]$ and $[\frac{1}{4}, \frac{1}{2}]$ (or one of $[\frac{1}{2}, \frac{3}{4}]$ and $[\frac{3}{4}, 1]$) contains infinitely many elements of $\{x_{n^{(1)}}\}$. Let this interval be I_2 and $\{x_{n^{(2)}}\}$ be the subsequence of $\{x_{n^{(1)}}\}$ such that

$$\{x_{n^{(2)}}\} \subseteq I_2$$

and the length of I_2 is $\frac{1}{4}$. This process can be continued inductively, we are able to construct a sequence of intervals

$$I_1 \supseteq I_2 \supseteq \cdots$$

and a sequence of infinite subsequences of the original sequence $\{x_n\}$

$$\{x_{n^{(1)}}\} \supseteq \{x_{n^{(2)}}\} \supseteq \cdots,$$

where the length of each I_k is $\frac{1}{2^k}$ and I_k contains $\{x_{n^{(k)}}\}$ for $k = 1, 2, \ldots$. Now we may pick one term y_k from each infinite subset $\{x_{n^{(k)}}\}$ and consider the subsequence $\{y_k\}$ of $\{x_n\}$.

We claim that $\{y_k\}$ is Cauchy. To this end, given $\epsilon > 0$, Theorem 2.1 (The Archimedean Property) implies that there exists a positive integer N such that $\frac{1}{2^N} < \epsilon$. Thus for this N, if $k, j > N$, then we have

$$|y_k - y_j| \le \max\left(\frac{1}{2^k}, \frac{1}{2^j}\right) < \frac{1}{2^N} < \epsilon.$$

This proves the claim. By Theorem 5.13(b), the subsequence $\{y_k\}$ is convergent and this completes the proof of the problem. ■

Problem 5.26

⋆ ⋆ *If the real sequence $\{x_n\}$ is unbounded, prove that there is an unbounded subsequence $\{x_{n_k}\}$.*

Proof. Since $\{x_n\}$ is unbounded, there exists a term x_{n_1} such that

$$|x_{n_1}| > 1.$$

Split the sequence $\{x_n\}$ into two subsequences $\{x_1, \ldots, x_{n_1-1}\}$ and $\{x_{n_1}, x_{n_1+1}, \ldots\}$. The first subsequence is obviously bounded, so the second subsequence must be unbounded. Thus there exists a term x_{n_2} in $\{x_{n_1}, x_{n_1+1}, \ldots\}$ such that

$$|x_{n_2}| > 2.$$

We may continue this way to obtain a subsequence $\{x_{n_k}\}$ satisfying the condition

$$|x_{n_k}| > k$$

for each positive integer k. By definition, the subsequence $\{x_{n_k}\}$ is unbounded. This finishes the proof of the problem. ■

Problem 5.27

$(\star)$ *Let X be a complete metric space. Prove that if a subset E of X is closed in X, then E is also a complete metric space with the same metric as X.*

Proof. Suppose that E is closed in X. Let $\{p_n\}$ be a Cauchy sequence of E. Since X is complete, there is a $p \in X$ such that $p_n \to p$. In other words, p is a limit point of $\{p_n\}$ by Theorem 5.2. Since E is closed in X, E must contain p. Since E is also a metric space (see §4.1.4), it follows from the definition in §5.1.4 that E is a complete metric space. Hence we have finished the proof of the problem.

■

Problem 5.28

$(\star)(\star)$ *Suppose that $\{p_n\}$ and $\{q_n\}$ are sequences in a metric space X with metric d. Let $\{p_n\}$ be Cauchy and $d(p_n, q_n) \to 0$ as $n \to \infty$. Prove that $\{q_n\}$ is Cauchy.*

Proof. Given that $\epsilon > 0$. Then there is a positive integer N_1 such that $m, n \geq N_1$ implies

$$d(p_m, p_n) < \frac{\epsilon}{3}. \tag{5.53}$$

Furthermore, since $d(p_n, q_n) \to 0$ as $n \to \infty$, there exists a positive integer N_2 such that

$$d(p_n, q_n) < \frac{\epsilon}{3} \tag{5.54}$$

for all $n \geq N_2$. Let $N = \max(N_1, N_2)$. Thus for $m, n \geq N$, we follow from the inequalities (5.53) and (5.54) that

$$d(q_m, q_n) \leq d(q_m, p_m) + d(p_m, p_n) + d(p_n, q_n) < \frac{\epsilon}{3} + \frac{\epsilon}{3} + \frac{\epsilon}{3} = \epsilon.$$

Hence $\{q_n\}$ is Cauchy and we complete the proof of the problem.

■

5.5 Recurrence Relations

Problem 5.29

$(\star)$ *Let $0 < A < 2$. Suppose that $\{p_n\}$ is a real sequence such that*

$$p_{n+1} = Ap_n + (1 - A)p_{n-1}, \tag{5.55}$$

where $n = 1, 2, \ldots$. Prove that $\{p_n\}$ converges.

Proof. Rewrite the relation (5.55) as

$$p_n - p_{n-1} = (A - 1)(p_{n-1} - p_{n-2})$$

for $n = 2, 3, \ldots$. Then it becomes

$$p_n - p_{n-1} = (A - 1)^{n-1}(p_1 - p_0) \tag{5.56}$$

for $n = 1, 2, \ldots$. By the formula (5.56), we get

$$
\begin{aligned}
p_n - p_0 &= (p_n - p_{n-1}) + (p_{n-1} - p_{n-2}) + \cdots + (p_1 - p_0) \\
&= [(A-1)^{n-1} + (A-1)^{n-2} + \cdots + (A-1) + 1](p_1 - p_0) \\
&= \frac{1 - (A-1)^n}{1 - (A-1)}(p_1 - p_0) \\
&= \frac{1 - (A-1)^n}{2 - A}(p_1 - p_0).
\end{aligned}
\tag{5.57}
$$

Since $0 < A < 2$, $-1 < A - 1 < 1$ and then Theorem 5.7(c) gives $\lim\limits_{n \to \infty}(1 - A)^n = 0$. By this, we obtain from the equation (5.57) that

$$\lim_{n \to \infty} p_n = p_0 + \lim_{n \to \infty} \frac{1 - (A-1)^n}{2 - A}(p_1 - p_0) = p_0 + \frac{1}{2 - A}(p_1 - p_0).$$

Hence $\{p_n\}$ converges and it finishes the proof of the problem. ∎

Problem 5.30

$(\star)$ *Suppose that $\{p_n\}$ is a real sequence defined by*

$$p_1 = 1 \quad \text{and} \quad p_{n+1} = \frac{1}{2}\left(p_n + \frac{1}{p_n}\right). \tag{5.58}$$

Find $\lim\limits_{n \to \infty} p_n$ if it exists.

Proof. It is clear that $p_n \geq 0$ for all $n = 1, 2, \ldots$. By the A.M. $\geq$ G.M., we can show further that

$$p_n \geq 1 \tag{5.59}$$

for all $n = 1, 2, \ldots$. Thus the sequence $\{p_n\}$ is bounded below by 1. By the inequality (5.59), we see that

$$p_{n+1} - p_n = \frac{1}{2}\left(p_n + \frac{1}{p_n}\right) - p_n = \frac{1}{2}\left(\frac{1 - p_n^2}{p_n}\right) \leq 0$$

for every $n = 1, 2, \ldots$. Therefore, the sequence $\{p_n\}$ is monotonically decreasing. By the Monotonic Convergence Theorem, we know that $\{p_n\}$ is convergent. Let

$$\lim_{n \to \infty} p_n = p$$

for some real p. By the recurrence relation (5.58), we gain that

$$
\begin{aligned}
p &= \frac{1}{2}\left(p + \frac{1}{p}\right) \\
p^2 &= 1 \\
p &= \pm 1.
\end{aligned}
$$

By the inequality (5.59), we must have $p \geq 1$, so $p = 1$. This ends the proof of the problem. ∎

Problem 5.31

$(\star)(\star)$ *Define the real sequence* $\{p_n\}$ *by* $p_1 = p_2 = 1$ *and*

$$p_n = \frac{p_{n-1}^2 + 2}{p_{n-2}} \tag{5.60}$$

for every positive integer $n \geq 3$. *Suppose that* $p_n = Ax^n + By^n$ *for some constants* A, B, x *and* y. *Prove that* $p_n \in \mathbb{N}$.

Proof. By the hypothesis $p_1 = p_2 = 1$, we have

$$Ax + By = 1 \quad \text{and} \quad Ax^2 + By^2 = 1. \tag{5.61}$$

Solving the equations (5.61), we have

$$Ax + \left(\frac{1 - Ax^2}{y^2}\right)y = 1$$

which implies that

$$A = \frac{y - 1}{x(y - x)}. \tag{5.62}$$

By substituting the constant (5.62) back to one of the equations (5.61), we gain

$$B = \frac{1 - x}{y(y - x)}. \tag{5.63}$$

Next, we put $p_n = Ax^n + By^n$ into the recurrence relation (5.60) to get

$$(Ax^n + By^n)(Ax^{n-2} + By^{n-2}) = (Ax^{n-1} + By^{n-1})^2 + 2$$
$$A^2x^{2n-2} + ABx^ny^{n-2} + ABx^{n-2}y^n + B^2y^{2n-2} = A^2x^{2n-2} + 2ABx^{n-1}y^{n-1} + B^2y^{2n-2} + 2.$$

After simplification, we have

$$ABx^ny^{n-2} + ABx^{n-2}y^n = 2ABx^{n-1}y^{n-1} + 2$$
$$ABx^ny^{n-2} - 2ABx^{n-1}y^{n-1} + ABx^{n-2}y^n = 2$$
$$ABx^{n-2}y^{n-2}(x^2 - 2xy + y^2) = 2$$
$$ABx^{n-2}y^{n-2}(x - y)^2 = 2. \tag{5.64}$$

By substituting the constants (5.62) and (5.63) into (5.64), we achieve that

$$(y - 1)(1 - x)(xy)^{n-3} = 2. \tag{5.65}$$

Since the equation (5.65) is valid for all $n = 3, 4, 5, \ldots$, we must have $xy = 1$ and then we have $(y - 1)(1 - x) = 2$ or equivalently

$$x + y = 4.$$

Now the conditions $xy = 1$ and $x + y = 4$ show that x and y are solutions of the equation

$$z^2 - 4z + 1 = 0,$$

so $x = 2 + \sqrt{3}$ and $y = 2 - \sqrt{3}$. Therefore, it follows from the constants (5.62) and (5.63) that

$$A = \frac{3 - \sqrt{3}}{6(2 + \sqrt{3})} \quad \text{and} \quad B = \frac{3 + \sqrt{3}}{6(2 - \sqrt{3})}.$$

Thus we conclude that

$$p_n = Ax^n + By^n = \frac{3 - \sqrt{3}}{6} \times (2 + \sqrt{3})^{n-1} + \frac{3 + \sqrt{3}}{6} \times (2 - \sqrt{3})^{n-1}$$

for every $n = 3, 4, \ldots$. Now it can be shown by induction that p_n is an integer for $n = 3, 4, \ldots$. Since $p_1 = p_2 = 1$, we establish that $p_n \in \mathbb{N}$ for all $n = 1, 2, \ldots$. This completes the proof of the problem. ∎

CHAPTER *6*

Series of Numbers

6.1 Fundamental Concepts

In this section, we summarize some basic results about series of real or complex numbers. The main references for the current topic are [3, Chap. 8], [5, §3.7, pp. 94 - 101], [13, §3.21 - §3.55, pp. 58 - 78] and [15, Chap. 7].

6.1.1 Definitions and Addition of Series

Suppose that $\{a_n\}$ is a sequence of real or complex numbers. We form a new sequence $\{s_n\}$, where

$$s_n = a_1 + a_2 + \cdots + a_n = \sum_{k=1}^{n} a_k \tag{6.1}$$

for each $n = 1, 2, \ldots$. Here we use the symbol

$$a_1 + a_2 + \cdots$$

to mean the **infinite series**

$$\sum_{n=1}^{\infty} a_n. \tag{6.2}$$

The numbers s_n are called the n-**th partial sums** of the series (6.2).

Definition 6.1. *The series (6.2) is said to* **converge** *or* **diverge** *according to the sequence $\{s_n\}$ is convergent or divergent. In the case of convergence, we write*

$$\sum_{n=1}^{\infty} a_n = s = \lim_{n \to \infty} s_n.$$

Here we call s the **sum** *of the series (6.2).*

The following result describes addition of convergent series.

Theorem 6.2. *Suppose that* $s = \sum_{n=1}^{\infty} a_n$ *and* $t = \sum_{n=1}^{\infty} b_n$ *are convergent series. Suppose, further, that* $\alpha, \beta \in \mathbb{C}$. *Then we obtain*

$$\sum_{n=1}^{\infty} (\alpha a_n \pm \beta b_n) = \alpha s \pm \beta t.$$

6.1.2 Tests for Convergence of Series

Many tests have been developed for determining the convergence of a given series. Some of them are given as follows:[a]

Theorem 6.3. *Suppose that* $a_n \geq 0$ *for* $n = 1, 2, \ldots$. *Then the series* $\sum a_n$ *converges if and only if the sequence* $\{s_n\}$ *is bounded, where* s_n *is the* n-*th partial sum (6.1).*

Theorem 6.4 (Cauchy Criterion). *The series* $\sum a_n$ *converges if and only if for every* $\epsilon > 0$, *there exists a positive integer* N *such that* $m \geq n \geq N$ *implies*

$$\left| \sum_{k=n}^{m} a_k \right| < \epsilon.$$

Theorem 6.5. *Suppose that* $\{a_n\}$ *is monotonically decreasing and bounded below by* 0. *Then the series* $\sum a_n$ *converges if and only if the series*

$$\sum 2^n a_{2^n}$$

converges.

The following three theorems are the famous **comparison test**, the **root test** and the **ratio test**.

Theorem 6.6 (Comparison Test). *Let* N *be a fixed positive integer.*

(a) *Suppose that* $|a_n| \leq b_n$ *for all* $n \geq N$. *If* $\sum b_n$ *converges, then* $\sum a_n$ *converges.*

(b) *Suppose that* $a_n \geq b_n \geq 0$ *for all* $n \geq N$. *If* $\sum b_n$ *is divergent, then* $\sum a_n$ *is divergent.*

Theorem 6.7 (Root Test). *Let* $\sum a_n$ *be a series of complex numbers and let*

$$\alpha = \limsup_{n \to \infty} \sqrt[n]{|a_n|}.$$

(a) *If* $\alpha < 1$, *then* $\sum a_n$ *converges.*

(b) *If* $\alpha > 1$, *then* $\sum a_n$ *diverges.*

(c) *If* $\alpha = 1$, *then no conclusion can be drawn.*

---

[a] Other useful tests can be found in [3, Theorems 8.10, 8.21 & 8.23, pp. 186, 190 & 191].

Theorem 6.8 (Ratio Test). *Let $\sum a_n$ be a series of complex numbers and let*

$$R = \limsup_{n \to \infty} \left| \frac{a_{n+1}}{a_n} \right|.$$

(a) If $R < 1$, then $\sum a_n$ converges.

(b) If $\left| \dfrac{a_{n+1}}{a_n} \right| \geq 1$ for all $n \geq N$, where N is a fixed positive integer, then $\sum a_n$ diverges.

Remark 6.1

We notice that the conclusions of the Root Test and the Ratio Test can be stronger. Firstly, we introduce the concept of **absolute convergence**: A series $\sum a_n$ is called **converges absolutely** if the series $\sum |a_n|$ converges. Then the absolute convergence of $\sum a_n$ implies the convergence of $\sum a_n$ and the conclusions of the Root Test and the Ratio Test can be changed from convergence to absolute convergence. Another importance property of absolute convergence will be given in §6.1.6.

6.1.3 Three Special Series

The following two results are very useful and important in many real problems. In fact, Theorem 6.9 (Geometric Series) can tell us the exact value of a certain type of series, called **geometric series**. Theorem 6.10 (p-series) is commonly applied with Theorem 6.6 (Comparison Test) to determine the convergence of other series.

Theorem 6.9 (Geometric Series). *If $-1 < x < 1$, then we have*

$$\sum_{n=0}^{\infty} x^n = \frac{1}{1-x}.$$

Theorem 6.10 (p-series). *The series*

$$\sum_{n=1}^{\infty} \frac{1}{n^p}$$

converges if $p > 1$ and diverges if $p \leq 1$.

The next result discusses about the convergence of the so-called **alternating series**. We have the precise definition first.

Definition 6.11. *If $a_n > 0$ for each $n = 1, 2, \ldots$, we call the series*

$$\sum_{n=1}^{\infty} (-1)^n a_n \tag{6.3}$$

*an **alternating series**.*

Theorem 6.12 (Alternating Series Test). *Suppose that $\{a_n\}$ is a monotonically decreasing sequence and $\lim\limits_{n \to \infty} a_n = 0$. Then the series (6.3) converges.*

▌6.1.4 Series in the form $\sum_{n=1}^{\infty} a_n b_n$

Sometimes we have to cope with series in the form

$$\sum_{n=1}^{\infty} a_n b_n. \tag{6.4}$$

The main tool for studying the convergence of the series (6.4) is the so-called **partial summation formula**.

Theorem 6.13. *Suppose that $\{a_n\}$ and $\{b_n\}$ are two sequences of complex numbers. Denote*

$$A_n = a_1 + \cdots + a_n.$$

Then we have the following identity

$$\sum_{k=1}^{n} a_k b_k = A_n b_{n+1} - \sum_{k=1}^{n} A_k (b_{k+1} - b_k).$$

By this tool, we can prove two famous results: **Dirichlet's Test** and **Abel's Test**.

Theorem 6.14 (Dirichlet's Test). *Suppose that $\sum a_n$ is a series of complex numbers whose partial sums $\{A_n\}$ is bounded. If $\{b_n\}$ is a monotonically decreasing sequence and $\lim_{n\to\infty} b_n = 0$, then the series (6.4) converges.*

Theorem 6.15 (Abel's Test). *Suppose that $\sum a_n$ is convergent and $\{b_n\}$ is a monotonic bounded sequence. Then the series (6.4) converges.*

▌6.1.5 Multiplication of Series

By Theorem 6.2, we see easily that two convergent series may be added term by term and the resulting series will converge to the sum of the two series. However, the situation is a little bit complicated when multiplication of series is considered. Let's define a product of two series first. It is called the **Cauchy product.**[b]

Definition 6.16. *Suppose that we have two series $\sum_{n=0}^{\infty} a_n$ and $\sum_{n=0}^{\infty} b_n$. Define*

$$c_n = \sum_{k=0}^{n} a_k b_{n-k},$$

*where $n = 0, 1, 2, \ldots$. Then the series $\sum_{n=0}^{\infty} c_n$ is the **Cauchy product** of the two series.*

Theorem 6.17 (Mertens' Theorem). *Suppose that $\sum_{n=0}^{\infty} a_n$ converges absolutely, $\sum_{n=0}^{\infty} a_n = A$ and $\sum_{n=0}^{\infty} b_n$ converges with sum B. Then the Cauchy product of the two series converges to AB, i.e.,*

$$\sum_{n=0}^{\infty} c_n = AB.$$

---

[b]There is another way to define multiplication of series which is known to be **Dirichlet product**.

6.1.6 Rearrangement of Series

It may happen that different rearrangements of a convergent series give different numbers. One example can be found in [13, Example 3.53, p. 76]. To make sure different rearrangements give the same sum, the series must converge absolutely:

Theorem 6.18. *If $\sum a_n$ is a series of complex numbers and it converges absolutely, then every rearrangement of $\sum a_n$ converges to the same sum.*

6.1.7 Power Series

Definition 6.19. *An infinite series of the form*

$$\sum_{n=0}^{\infty} c_n z^n \tag{6.5}$$

*is called a **power series**. Here all c_n and z are complex numbers. The numbers c_n are called the **coefficients** of the series (6.5).*

It is well-known that with every power series there is associated a circle, namely the **circle of convergence**, such that the series (6.5) converges for every z in the interior of the circle and diverges if z is outside the circle. The following theorem provides us an easy method to establish the circle of convergence.

Theorem 6.20. *Given a power series in the form (6.5). Take*

$$\alpha = \limsup_{n \to \infty} \sqrt[n]{|c_n|} \quad and \quad R = \frac{1}{\alpha}.$$

*Then the series (6.5) converges if $|z| < R$ and diverges if $|z| > R$. The number R is called the **radius of convergence** of the power series (6.5).*

6.2 Convergence of Series of Nonnegative Terms

Unless otherwise specified the terms in the series are assumed to be nonnegative.

Problem 6.1

(⋆) Prove that $\displaystyle\sum_{n=1}^{\infty} \frac{n}{n^4 - n^2 + 1}$ converges.

Proof. By simple algebra, we can show that

$$0 < \frac{n}{n^4 - n^2 + 1} < \frac{2}{n^3}$$

because $(n^2 - 1)^2 + 1 > 0$ for every positive integer n. By Theorem 6.10 (p-series),

$$\sum_{n=1}^{\infty} \frac{1}{n^3}$$

converges. Then Theorem 6.6 (Comparison Test) implies that

$$\sum_{n=1}^{\infty} \frac{n}{n^4 - n^2 + 1}$$

converges which completes the proof of the problem.   ∎

Problem 6.2

(⋆) (⋆) Let $a_n \geq 0$ for $n = 1, 2, \ldots$. Suppose that $\displaystyle\sum_{n=1}^{\infty} n a_n$ converges. Prove that $\displaystyle\sum_{n=1}^{\infty} a_n$ converges.

Proof. Let $s_n = a_1 + 2a_2 + \cdots + n a_n$ for $n = 1, 2, \ldots$. Since $\displaystyle\sum_{n=1}^{\infty} n a_n$ converges, Theorem 6.3 implies that $\{s_n\}$ is bounded, i.e., there exists a positive constant M such that

$$|s_n| \leq M$$

for all positive integers n. For $k = 1, 2, \ldots$, we note that

$$a_k = \frac{s_k - s_{k-1}}{k}. \tag{6.6}$$

Given that $\epsilon > 0$, for positive integers m, n with $m > n > \frac{2M}{\epsilon}$, we deduce from the equation (6.6) that

$$
\begin{aligned}
\left| \sum_{k=n}^{m} a_k \right| &= \left| \frac{s_n - s_{n-1}}{n} + \frac{s_{n+1} - s_n}{n+1} + \cdots + \frac{s_m - s_{m-1}}{m} \right| \\
&= \left| -\frac{s_{n-1}}{n} + \left(\frac{1}{n} - \frac{1}{n+1} \right) s_n + \cdots + \left(\frac{1}{m-1} - \frac{1}{m} \right) s_{m-1} + \frac{s_m}{m} \right| \\
&\leq M \left[\frac{1}{n} + \left(\frac{1}{n} - \frac{1}{n+1} \right) + \left(\frac{1}{n+1} - \frac{1}{n+2} \right) + \cdots + \left(\frac{1}{m-1} - \frac{1}{m} \right) + \frac{1}{m} \right] \\
&= \frac{2M}{n} \\
&< \epsilon.
\end{aligned}
$$

By Theorem 6.4 (Cauchy Criterion), the series $\displaystyle\sum_{n=1}^{\infty} a_n$ converges. This completes the proof of the problem.   ∎

Problem 6.3

$(\star)$ *Suppose that $a_1, a_2, \ldots$ are complex numbers and $\displaystyle\sum_{n=1}^{\infty} a_n$ converges. Prove that*

$$\lim_{n \to \infty} a_n = 0. \tag{6.7}$$

Proof. If we take $m = n$ in Theorem 6.4 (Cauchy Criterion), then we have

$$|a_n| < \epsilon$$

for all $n \geq N$. Thus it follows from the definition that the limit (6.7) holds. We complete the proof of the problem. ∎

Problem 6.4

$(\star)(\star)$ *Prove that $\displaystyle\sum_{n=1}^{\infty} \frac{1}{n}$ is divergent without using Theorem 6.10 (p-series).*

Proof. Let $s_n = 1 + \frac{1}{2} + \cdots + \frac{1}{n}$ for each positive integer n. By the fact $e^x > 1 + x > 0$ for $x > 0$, we have

$$\begin{aligned}
e^{s_n} &= \exp\left(1 + \frac{1}{2} + \cdots + \frac{1}{n}\right) \\
&= e^1 \times e^{\frac{1}{2}} \times \cdots \times e^{\frac{1}{n}} \\
&> (1 + 1)\left(1 + \frac{1}{2}\right) \times \cdots \times \left(1 + \frac{1}{n}\right) \\
&= 2 \times \frac{3}{2} \times \cdots \times \frac{n}{n-1} \times \frac{n+1}{n} \\
&= n + 1.
\end{aligned}$$

Thus $\{e^{s_n}\}$ is unbounded and hence $\{s_n\}$ is also unbounded. By Theorem 6.3, the series

$$\sum_{n=1}^{\infty} \frac{1}{n}$$

is divergent. This completes the proof of the problem. ∎

Remark 6.2

The series in Problem 6.4 is called the **harmonic series**. There are many proofs of Problem 6.4 and Theorem 6.10 (p-series) is clearly one of them.

> **Problem 6.5**
>
> ⊛ *Suppose that $a_n \geq 0$ for $n = 1, 2, \ldots$ and $\displaystyle\sum_{n=1}^{\infty} a_n$ converges. Prove that the series*
>
> $$\sum_{n=1}^{\infty} a_n^2$$
>
> *converges. Is the converse true?*

Proof. Since $\displaystyle\sum_{n=1}^{\infty} a_n$ converges, it follows from Problem 6.3 that $a_n \to 0$ as $n \to \infty$. Therefore, there exists a positive constant M such that

$$0 \leq a_n < M$$

for $n = 1, 2, \ldots$. Then we have

$$0 \leq a_n^2 < M a_n$$

for $n = 1, 2, \ldots$. By Theorem 6.2, the series $\displaystyle\sum_{n=1}^{\infty} M a_n$ converges. By Theorem 6.6 (Comparison Test), the series

$$\sum_{n=1}^{\infty} a_n^2$$

converges.

The converse is false. For examples, the series $\displaystyle\sum_{n=1}^{\infty} \frac{1}{n^2}$ converges by Theorem 6.10 (p-series). However, Theorem 6.10 (p-series) or Problem 6.4 shows that the series $\displaystyle\sum_{n=1}^{\infty} \frac{1}{n}$ is divergent. Thus we have completed the proof of the problem.   ∎

> **Problem 6.6**
>
> ⊛ *Prove that the series $\displaystyle\sum_{n=2}^{\infty} \frac{1}{n(\log n)}$ diverges.*

Proof. Since the sequence $\{\log n\}$ is increasing, the sequence $\{\frac{1}{n(\log n)}\}$ is decreasing and it is clear that $\{\frac{1}{n(\log n)}\}$ is bounded below by 0 for $n = 2, 3, \ldots$. Consider

$$\sum_{k=2}^{n} 2^k \times \frac{1}{2^k(\log 2^k)} = \frac{1}{\log 2} \sum_{k=2}^{n} \frac{1}{k}.$$

By Theorem 6.10 (p-series),

$$\sum_{n=2}^{\infty} \frac{1}{n}$$

diverges. Thus it follows from Theorem 6.6 (Comparison Test) that

$$\sum_{n=2}^{\infty} \frac{1}{n(\log n)}$$

is divergent. Hence we finish the proof of the problem. ∎

Problem 6.7

(⋆) Prove that the series $\sum_{n=1}^{\infty} \frac{n^n}{7^{n^2}}$ converges.

Proof. Since

$$\limsup_{n\to\infty} \sqrt[n]{\frac{n^n}{7^{n^2}}} = \lim_{n\to\infty} \frac{n}{7^n} = 0,$$

we know from Theorem 6.7 (Root Test) that the series converges. We have completed the proof of the problem. ∎

Problem 6.8

(⋆) Prove that both series

$$\sum_{n=1}^{\infty} \frac{1}{n!} \quad \text{and} \quad \sum_{n=1}^{\infty} \frac{1}{n^n} \tag{6.8}$$

converge.

Proof. Put $a_n = \frac{1}{n!}$ so that

$$\frac{a_{n+1}}{a_n} = \frac{1}{n+1}$$

which gives

$$R = \limsup_{n\to\infty} \left| \frac{a_{n+1}}{a_n} \right| = \lim_{n\to\infty} \frac{1}{n+1} = 0.$$

By Theorem 6.8 (Ratio Test), we see that the first series in (6.8) converges.

Since $\log k \leq \log n$ for all $k = 1, 2, \ldots, n$, we must have

$$0 \leq \log(n!) = \sum_{k=1}^{n} \log k \leq \sum_{k=1}^{n} \log n = \log(n^n) \tag{6.9}$$

for every positive integer n. By the inequalities (6.9), we deduce that $n! \leq n^n$ for each $n = 1, 2, \ldots$ and then

$$0 < \frac{1}{n^n} \leq \frac{1}{n!}.$$

By our first assertion and Theorem 6.6 (Comparison Test), we conclude that

$$\sum_{n=1}^{\infty} \frac{1}{n^n}$$

also converges, completing the proof of the problem. ∎

Problem 6.9

$\star$ $\star$ *Discuss the convergence of the series*

$$\sum_{n=2}^{\infty} \frac{1}{(\log n)^{\log n}} \quad \text{and} \quad \sum_{n=3}^{\infty} \frac{1}{(\log \log n)^{\log \log n}}. \tag{6.10}$$

Proof. Since $a^{\log b} = b^{\log a}$ holds for $a > 0$ and $b > 0$, we have

$$(\log n)^{\log n} = n^{\log \log n} \tag{6.11}$$

for every positive integer $n \geq 2$. Furthermore, when $n \geq e^{e^2}$, we have $\log \log n \geq 2$ so it follows from the expression (6.11) that

$$(\log n)^{\log n} = n^{\log \log n} \geq n^2 > 0$$

or equivalently,

$$0 < \frac{1}{(\log n)^{\log n}} \leq \frac{1}{n^2}$$

for $n \geq e^{e^2}$. By Theorem 6.10 (p-series), we know that $\displaystyle\sum_{n \geq e^{e^2}}^{\infty} \frac{1}{n^2}$ converges. Thus Theorem 6.6 (Comparison Test) shows that

$$\sum_{n \geq e^{e^2}}^{\infty} \frac{1}{(\log n)^{\log n}}$$

is convergent and the convergence of the first series in (6.10) follows from this fact immediately.

We claim that the second series in (6.10) is divergent. To this end, we first note that for $n \geq 3$, we always have $\log n > \log \log n > 0$ so that

$$\frac{1}{(\log \log n)^{\log \log n}} > \frac{1}{(\log n)^{\log \log n}} \tag{6.12}$$

for $n \geq 3$. Furthermore, we follow from the fact $a^b = e^{b \log a}$ for $a > 0$ and $b \in \mathbb{R}$ that

$$(\log n)^{\log \log n} = e^{(\log \log n) \times (\log \log n)}. \tag{6.13}$$

By the fact[c] that $\frac{x}{e^x} \to 0$ as $x \to +\infty$, we obtain

$$\frac{x}{e^{\frac{x}{2}}} \to 0$$

as $x \to +\infty$. In other words, for large x, we must have

$$0 < x < e^{\frac{x}{2}}$$

or equivalently

$$0 < 2 \log x < x. \tag{6.14}$$

---

[c]See [13, Theorem 8.6(f), p. 180].

By substituting $x = \log \log n$ for large enough n into the inequalities (6.14), we see that

$$0 < 2 \log \log \log n < \log \log n$$

and they imply that

$$0 < (\log \log n)^2 < \log n. \tag{6.15}$$

Combining the inequalities (6.15) and the equation (6.13), we can show that

$$\frac{1}{(\log n)^{\log \log n}} = \frac{1}{e^{(\log \log n)^2}} > \frac{1}{e^{\log n}} = \frac{1}{n}. \tag{6.16}$$

Next, we achieve from the inequalities (6.12) and (6.16) that

$$\frac{1}{(\log \log n)^{\log \log n}} > \frac{1}{n}.$$

Finally, by using Theorem 6.10 (p-series) and then Theorem 6.6 (Comparison Test), we conclude that

$$\sum_{n=3}^{\infty} \frac{1}{(\log \log n)^{\log \log n}}$$

is divergent. This completes the proof of the problem. ∎

Problem 6.10

⋆ ⋆ ⋆ *Suppose that, for $n = 1, 2, \ldots$, we have*

$$0 < a_n \le \sum_{k=n+1}^{\infty} a_k \quad \text{and} \quad \sum_{n=1}^{\infty} a_n = A \tag{6.17}$$

for some constant A. Prove that there is either a finite subsequence $\{n_1, n_2, \ldots, n_j\}$ such that $a_{n_1} + \cdots + a_{n_j} = p$ or an infinite subsequence $\{n_j\}$ such that $\sum_{j=1}^{\infty} a_{n_j} = p$, where $0 < p < 1$.

Proof. Fix $p \in (0, 1)$. Let n_1 be the *least* positive integer such that

$$a_{n_1} \le p. \tag{6.18}$$

Such an n_1 must exist, otherwise,

$$\sum_{n=1}^{m} a_n > \sum_{n=1}^{m} p = mp > A$$

for some positive integer m, a contradiction. Furthermore, the inequality (6.18) implies that

$$\sum_{k=n_1}^{\infty} a_k \ge p.$$

Otherwise, the first hypothesis in (6.17) shows that

$$a_{n_1-1} \leq \sum_{k=n_1}^{\infty} a_k < p$$

which obviously contradicts the definition of n_1 in (6.18).

If $\sum_{k=n_1}^{\infty} a_k = p$, then an infinite subsequence $\{n_1, n_1+1, n_1+2, \ldots\}$ exists. Otherwise, we have $\sum_{k=n_1}^{\infty} a_k > p$ and let n_2 be the *greatest* integer such that

$$\sum_{k=n_1}^{n_2} a_k \leq p. \tag{6.19}$$

The existence of such an n_2 is guaranteed by the condition (6.18). If the equality holds, then a finite subsequence $\{n_1, n_1 + 1, \ldots, n_2\}$ exists. Otherwise, we have

$$\sum_{k=n_1}^{n_2} a_k < p$$

and then there exists an integer n_3 which is the *least* integer greater than n_2 such that

$$\sum_{k=n_1}^{n_2} a_k + a_{n_3} \leq p. \tag{6.20}$$

We remark that $n_3 > n_2 + 1$, otherwise, $n_3 = n_2 + 1$ which contradicts the definition of n_2 given by the inequality (6.19). In addition, we must have

$$\sum_{k=n_1}^{n_2} a_k + \sum_{k=n_3}^{\infty} a_k \geq p.$$

Otherwise, the first hypothesis in (6.17) implies that

$$p > \sum_{k=n_1}^{n_2} a_k + \sum_{k=n_3}^{\infty} a_k \geq \sum_{k=n_1}^{n_2} a_k + a_{n_3-1}$$

which obviously contradicts the definition of n_3 given by the inequality (6.20). If the equality holds, then an infinite subsequence $\{n_1, \ldots, n_2, n_3, n_3 + 1, \ldots\}$ exists. Otherwise, we have

$$\sum_{k=n_1}^{n_2} a_k + \sum_{k=n_3}^{\infty} a_k > p$$

and then we let n_4 be the *greatest* integer such that

$$\sum_{k=n_1}^{n_2} a_k + \sum_{k=n_3}^{n_4} a_k \leq p.$$

Similar to n_2, the existence of n_4 is guaranteed by the condition (6.20). If the equality holds, then a finite subsequence $\{n_1, \ldots, n_2, n_3, \ldots, n_4\}$ exists.

Now we may see that the above process can be continued. If the process terminates after finitely many steps, then we have either

$$a_{n_1} + \cdots + a_{n_j} = p \quad \text{or} \quad \sum_{j=1}^{\infty} a_{n_j} = p.$$

If the process does not terminate, then we must have

$$\sum_{k=n_1}^{n_2} a_k + \cdots + \sum_{k=n_{2j-1}}^{n_{2j}} a_k + \sum_{k=n_{2j}+1}^{\infty} a_k > p \tag{6.21}$$

and

$$\sum_{k=n_1}^{n_2} a_k + \cdots + \sum_{k=n_{2j-1}}^{n_{2j}} a_k + \sum_{k=n_{2j}+1}^{n_{2j+2}} a_k < p \tag{6.22}$$

where $n_{2j+1} > n_{2j} + 1$ for every positive integer j. Combining the inequalities (6.21) and (6.22) to get

$$0 < p - \left[\sum_{k=n_1}^{n_2} a_k + \cdots + \sum_{k=n_{2j-1}}^{n_{2j}} a_k \right] < \sum_{k=n_{2j}+1}^{\infty} a_k.$$

Since $n_{2j+1} \to +\infty$ as $j \to +\infty$, the second hypothesis in (6.17) and Problem 6.3 ensure that

$$\lim_{j \to \infty} \sum_{k=n_{2j}+1}^{\infty} a_k = 0.$$

By this and Theorem 5.6 (Squeeze Theorem for Convergent Sequences), an infinite subsequence $\{n_1, n_2, n_3, n_4, \ldots\}$ (after renaming) exists such that

$$\sum_{j=1}^{\infty} a_{n_j} = p.$$

This completes the proof of the problem. ∎

6.3 Alternating Series and Absolute Convergence

Problem 6.11

(★) *Determine the convergence of the series*

$$\sum_{n=1}^{\infty} \frac{\cos n\pi}{n^{\frac{4}{5}}}. \tag{6.23}$$

Proof. Recall the fact that

$$\cos n\pi = (-1)^n$$

for $n = 1, 2, \ldots$. Then we have

$$\sum_{n=1}^{\infty} \frac{\cos n\pi}{n^{\frac{4}{5}}} = \sum_{n=1}^{\infty}(-1)^n \frac{1}{n^{\frac{4}{5}}}$$

which is an alternating series. It is clear that the sequence $\{n^{-\frac{4}{5}}\}$ is monotonically decreasing and

$$\lim_{n \to \infty} \frac{1}{n^{\frac{4}{5}}} = 0.$$

By Theorem 6.12 (Alternating Series Test), the series (6.23) is convergent and this completes the proof of the problem. ∎

Problem 6.12

$(\star)$ *Discuss the convergence of the series*

$$\sum_{n=1}^{\infty} \frac{(-1)^{n-1}}{n^p}. \tag{6.24}$$

Proof. If $p < 0$, then $n^{-p} \to +\infty$ as $n \to +\infty$. By Problem 6.3, the series (6.24) is divergent in this case. Similarly, the series (6.24) is also divergent if $p = 0$. When $p > 0$, we know that

$$\frac{1}{n^p} > \frac{1}{(n+1)^p}$$

for $n = 1, 2, \ldots$ so that $\{n^{-p}\}$ is monotonically decreasing. Besides, we have $\frac{1}{n^p} \to 0$ as $n \to \infty$. Thus Theorem 6.12 (Alternating Series Test) implies that the series (6.24) converges in this case. This ends the proof of the problem. ∎

Problem 6.13

$(\star)$ *In Remark 6.1, we know that absolute convergence implies convergence. Is the converse true?*

Proof. The converse is false. For example, the series

$$\sum_{n=1}^{\infty}(-1)^n \frac{1}{n}$$

converges by Theorem 6.12 (Alternating Series Test), but $\displaystyle\sum_{n=1}^{\infty} \frac{1}{n}$ is divergent by Problem 6.4. This finishes the proof of the problem. ∎

Problem 6.14

(⋆) Let a_n be real numbers for $n = 1, 2, \ldots$. Suppose that $\displaystyle\sum_{n=1}^{\infty} a_n$ converges absolutely. Prove that $\displaystyle\sum_{n=1}^{\infty} a_n^2$ converges absolutely.

Proof. Since $\displaystyle\sum_{n=1}^{\infty} a_n$ converges, Problem 6.3 implies that $a_n \to 0$ as $n \to \infty$. Thus there exists a positive integer N such that $n \geq N$ implies

$$|a_n| < 1.$$

Then we must have

$$0 \leq a_n^2 \leq |a_n|$$

for $n \geq N$. Since $\displaystyle\sum_{n=1}^{\infty} |a_n|$ converges, Theorem 6.6 (Comparison Test) implies that

$$\sum_{n=N}^{\infty} a_n^2 \tag{6.25}$$

converges. Since $\displaystyle\sum_{n=1}^{N-1} a_n^2$ is finite, we follow from this and the convergence of (6.25) that

$$\sum_{n=1}^{\infty} a_n^2$$

converges. Since $|a_n^2| = a_n^2$ for each $n = 1, 2, \ldots$, the series actually converges absolutely. Hence this completes the proof of the problem. ∎

Problem 6.15

(⋆)(⋆) Let

$$\sum_{n=3}^{\infty} \frac{a^n}{n^b (\log n)^c}, \tag{6.26}$$

where $a, b, c \in \mathbb{R}$. Prove that the series (6.26) converges absolutely if $|a| < 1$.

Proof. If $a = 0$, then each term in the series (6.26) is zero, so it converges absolutely. Suppose that $a \neq 0$. Now we have

$$\limsup_{n \to \infty} \left| \left| \frac{a^{n+1}}{(n+1)^b [\log(n+1)]^c} \right| \times \left| \frac{n^b (\log n)^c}{a^n} \right| \right| = |a| \lim_{n \to \infty} \left\{ \left(\frac{n}{n+1} \right)^b \times \left[\frac{\log n}{\log(n+1)} \right]^c \right\}. \tag{6.27}$$

Since
$$\lim_{n\to\infty} \frac{n}{n+1} = 1 \quad \text{and} \quad \lim_{n\to\infty} \frac{\log n}{\log(n+1)} = 1,$$

we deduce from the expression (6.27) that

$$\limsup_{n\to\infty} \left| \left| \frac{a^{n+1}}{(n+1)^b[\log(n+1)]^c} \right| \times \left| \frac{n^b(\log n)^c}{a^n} \right| \right| = |a|.$$

Therefore, we follow from Theorem 6.8 (Ratio Test) that the series (6.26) converges absolutely when $|a| < 1$ and $a = 0$. As we have shown in the previously that it converges absolutely when $a = 0$, we gain our desired result that it converges absolutely if $|a| < 1$. We have finished the proof of the problem. ∎

6.4 The Series $\sum_{n=1}^{\infty} a_n b_n$ and Multiplication of Series

Problem 6.16

⋆ ⋆ *Suppose that* $\displaystyle\sum_{n=1}^{\infty} a_n^2$ *and* $\displaystyle\sum_{n=1}^{\infty} b_n^2$ *converge,* $a_n \geq 0$ *and* $b_n \geq 0$ *for every* $n = 1, 2, \ldots$. *Prove that*

$$\sum_{n=1}^{\infty} a_n b_n$$

converges absolutely.

Proof. Let
$$\sum_{n=1}^{\infty} a_n^2 = A \quad \text{and} \quad \sum_{n=1}^{\infty} b_n^2 = B$$

respectively. Recall the **Schwarz inequality** that if $a_1, a_2, \ldots, a_n$ and $b_1, b_2, \ldots, b_n$ are complex numbers, then we have

$$\left| \sum_{k=1}^{n} a_k \bar{b}_k \right|^2 \leq \left(\sum_{k=1}^{n} |a_k|^2 \right) \times \left(\sum_{k=1}^{n} |b_k|^2 \right). \tag{6.28}$$

Since a_n and b_n are nonnegative for every $n \in \mathbb{N}$, we have

$$\left| \sum_{k=1}^{n} a_k b_k \right| = \sum_{k=1}^{n} a_k b_k = \sum_{k=1}^{n} |a_k b_k| \tag{6.29}$$

for all $n \in \mathbb{N}$. Now we substitute the expression (6.29) into the inequality (6.28) to obtain

$$\sum_{k=1}^{n} |a_k b_k| \leq \left(\sum_{k=1}^{n} a_k^2 \right)^{\frac{1}{2}} \times \left(\sum_{k=1}^{n} b_k^2 \right)^{\frac{1}{2}} < \sqrt{AB}$$

for every $n \in \mathbb{N}$. Thus the partial sums of the series $\displaystyle\sum_{n=1}^{\infty} |a_n b_n|$ is bounded. Hence it follows from

Theorem 6.3 that $\displaystyle\sum_{n=1}^{\infty} |a_n b_n|$ converges, i.e., $\displaystyle\sum_{n=1}^{\infty} a_n b_n$ converges absolutely by definition. This completes the proof of the problem. ∎

Problem 6.17

(★) *Suppose that $0 < \theta < 1$. Prove that the series $\displaystyle\sum_{n=1}^{\infty} \frac{e^{2\pi i n \theta}}{n}$ converges.*

Proof. Let $a_n = e^{2\pi i n \theta}$ and $b_n = \frac{1}{n}$ for each $n = 1, 2, \ldots$. It is clear that $\{b_n\}$ is monotonically decreasing and

$$\lim_{n \to \infty} b_n = \lim_{n \to \infty} \frac{1}{n} = 0.$$

Since $0 < \theta < 1$, we have $e^{2\pi i \theta} \neq 1$. Thus we have

$$A_n = \sum_{k=1}^{n} a_k = \sum_{k=1}^{n} e^{2\pi i k \theta} = \frac{e^{2\pi i \theta} - e^{2\pi i (n+1)\theta}}{1 - e^{2\pi i \theta}}. \tag{6.30}$$

Fix the θ. Since $|e^{2\pi i \theta}| = 1$, we deduce from the expression (6.30) that

$$|A_n| \leq \frac{2}{|1 - e^{2\pi i \theta}|}$$

which means that $\{A_n\}$ is bounded. Hence we apply Theorem 6.14 (Dirichlet's Test) to conclude that the series is convergent and we have finished the proof of the problem. ∎

Problem 6.18

(★) Prove that the series

$$\sum_{n=1}^{\infty} \frac{n^4 \sin(\frac{1}{n^3})}{e^n (n+1)} \tag{6.31}$$

converges.

Proof. For each $n = 1, 2, \ldots$, let

$$a_n = \frac{n^4}{e^n (n+1)} \quad \text{and} \quad b_n = \sin\left(\frac{1}{n^3}\right).$$

Consider

$$\limsup_{n \to \infty} \left| \frac{a_{n+1}}{a_n} \right| = \lim_{n \to \infty} \frac{(n+1)^4}{e^{n+1}(n+2)} \times \frac{e^n (n+1)}{n^4}$$

$$= \lim_{n \to \infty} \frac{1}{e} \times \left(\frac{n+1}{n}\right)^4 \times \left(\frac{n+1}{n+2}\right)$$

$$= \frac{1}{e}.$$

By Theorem 6.8, we see that the series $\sum\limits_{n=1}^{\infty} a_n$ converges. It is clear that $\{b_n\}$ is bounded by

1. Furthermore, since $\frac{1}{n^3} \to 0$ as $n \to \infty$ and $\sin x$ is decreasing on the interval $[0, \frac{\pi}{2}]$, $\{b_n\}$ is monotonically decreasing. Hence we follow from Theorem 6.15 (Abel's Test) that the series (6.31) is convergent. This completes the proof of the problem.   ∎

Problem 6.19

⋆ ⋆ *Suppose that* $\sum\limits_{n=0}^{\infty} a_n$ *and* $\sum\limits_{n=0}^{\infty} b_n$ *converge absolutely. Prove that their Cauchy product also converges absolutely.*

Proof. We have to prove that

$$C_n = \sum_{n=0}^{\infty} |c_n|$$

converges. Since $\sum\limits_{n=0}^{\infty} a_n$ and $\sum\limits_{n=0}^{\infty} b_n$ converge absolutely, we let

$$A = \sum_{n=0}^{\infty} |a_n|, \quad A_n = \sum_{k=0}^{n} |a_k|, \quad B = \sum_{n=0}^{\infty} |b_n| \quad \text{and} \quad B_n = \sum_{k=0}^{n} |b_k|.$$

Now for all $n \geq 0$, $|a_n|$ and $|b_n|$ are nonnegative terms of A_n and B_n respectively. This implies that

$$0 \leq A_n \leq A \quad \text{and} \quad 0 \leq B_n \leq B$$

for all $n \geq 0$. Therefore, we achieve that

$$
\begin{aligned}
|C_n| &= \sum_{n=0}^{\infty} |c_n| \\
&\leq |a_0||b_0| + (|a_0||b_1| + |a_1||b_0|) + \cdots + (|a_0||b_n| + |a_1||b_{n-1}| + \cdots + |a_n||b_0|) \\
&= |a_0|B_n + |a_1|B_{n-1} + \cdots + |a_n|B_0 \\
&\leq (|a_0| + |a_1| + \cdots + |a_n|)B \\
&= A_n B \\
&\leq AB.
\end{aligned}
$$

Hence $\{C_n\}$ is bounded and it follows from Theorem 6.3 that $\sum\limits_{n=1}^{\infty} |c_n|$ converges, i.e., the series

$$\sum_{n=0}^{\infty} c_n$$

converges absolutely.   ∎

> **Problem 6.20**
>
> $\star$ $\star$ *Does the Cauchy product of the series* $\displaystyle\sum_{n=0}^{\infty} \frac{(-1)^{n+1}}{\sqrt{n+1}}$ *with itself converge?*

Proof. Let $a_n = \frac{(-1)^{n+1}}{\sqrt{n+1}}$. By Definition 6.16, we have

$$
\begin{aligned}
c_n &= \sum_{k=0}^{n} a_k a_{n-k} \\
&= \sum_{k=0}^{n} \frac{(-1)^{k+1}}{\sqrt{k+1}} \times \frac{(-1)^{n-k+1}}{\sqrt{n-k+1}} \\
&= (-1)^n \sum_{k=0}^{n} \frac{1}{\sqrt{k+1}\sqrt{n-k+1}}.
\end{aligned}
\tag{6.32}
$$

Since $k+1$ and $n-k+1$ are nonnegative for all $n, k = 0, 1, \ldots$ with $0 \le k \le n$, the A.M. $\ge$ G.M. implies that

$$
\sqrt{(k+1)(n-k+1)} \le \frac{k+1+n-k+1}{2} = \frac{n+2}{2}.
\tag{6.33}
$$

Now, by combining the expression (6.32) and the inequality (6.33), we derive that

$$
|c_n| = \sum_{k=0}^{n} \frac{1}{\sqrt{k+1}\sqrt{n-k+1}} \ge \sum_{k=0}^{n} \frac{2}{n+2} = \frac{2n+2}{n+2}.
$$

Since $\frac{2n+2}{n+2} \to 2$ as $n \to \infty$, it means that there exists a positive integer N such that

$$
|c_n| > 1
$$

for all $n \ge N$. Thus it follows from Problem 6.3 that the series

$$
\sum_{n=0}^{\infty} c_n
$$

is divergent. Hence we complete the proof of the problem.   ■

> **Remark 6.3**
>
> It is clear that
> $$
> \sum_{n=0}^{\infty} \frac{(-1)^{n+1}}{\sqrt{n+1}}
> $$
> converges by Theorem 6.12 (Alternating Series Test), Problem 6.20 tells us that absolute convergence of one of the two series in Theorem 6.17 (Mertens' Theorem) cannot be dropped.

6.5 Power Series

Problem 6.21

($\star$) *Determine the radius of convergence of each of the following series:*

(a) $\displaystyle\sum_{n=1}^{\infty}(\log 2n)z^{n}$.

(b) $\displaystyle\sum_{n=1}^{\infty}\frac{z^{n}}{(2n)!}$.

Proof.

(a) Since $1 \leq \sqrt[n]{\log 2n} \leq \sqrt[n]{n}$ for $n \geq 2$, we deduce from Theorems 5.7(b) and 6.6 (Comparison Test) that

$$\lim_{n\to\infty} \sqrt[n]{\log 2n} = 1.$$

By Theorem 6.20, the radius of convergence of the power series is 1.

(b) For $n = 1, 2, \ldots$, we have

$$0 < \frac{1}{(2n)!} \leq \frac{1}{n!}. \tag{6.34}$$

By Problem 5.14, we must have

$$\lim_{n\to\infty} \sqrt[n]{n!} = +\infty. \tag{6.35}$$

Now we apply Theorem 6.6 (Comparison Test) and the limit (6.35) to the inequalities (6.34) to conclude that

$$\lim_{n\to\infty} \frac{1}{\sqrt[n]{(2n)!}} = 0.$$

By Theorem 6.20, the radius of convergence of the power series is $+\infty$.

∎

Problem 6.22

($\star$)($\star$) *Suppose that the radius of convergence of the power series $\displaystyle\sum_{n=0}^{\infty} c_n z^n$ is 3. Prove that 1 is the radius of convergence of the power series*

$$\sum_{n=0}^{\infty} c_n z^{n^3}. \tag{6.36}$$

Proof. By the hypothesis, we know that

$$\limsup_{n\to\infty} \sqrt[n]{|c_n|} = \frac{1}{3}.$$

We consider

$$\limsup_{n\to\infty} \sqrt[n]{|c_n z^{n^3}|} = \limsup_{n\to\infty} (\sqrt[n]{|c_n|} \times |z|^{n^2}).$$

By Remark 5.1, we obtain

$$\liminf_{n\to\infty} |z|^{n^2} \times \limsup_{n\to\infty} \sqrt[n]{|c_n|} \le \limsup_{n\to\infty} (\sqrt[n]{|c_n|} \times |z|^{n^2}) \le \limsup_{n\to\infty} \sqrt[n]{|c_n|} \times \limsup_{n\to\infty} |z|^{n^2}. \quad (6.37)$$

By Theorem 5.10, we have

$$\liminf_{n\to\infty} |z|^{n^2} = \limsup_{n\to\infty} |z|^{n^2} = \lim_{n\to\infty} |z|^{n^2}. \quad (6.38)$$

Thus we follow from the inequalities (6.37) and (6.38) that

$$\limsup_{n\to\infty} \sqrt[n]{|c_n z^{n^3}|} = \lim_{n\to\infty} |z|^{n^2} \times \limsup_{n\to\infty} \sqrt[n]{|c_n|} = \frac{1}{3} \lim_{n\to\infty} |z|^{n^2}.$$

By Theorem 6.7 (Root Test), the power series (6.36) converges if

$$\frac{1}{3} \lim_{n\to\infty} |z|^{n^2} < 1. \quad (6.39)$$

It is obvious that the inequality (6.39) holds if $|z| < 1$. However, if $|z| > 1$, then we must have

$$\lim_{n\to\infty} |z|^{n^2} = +\infty$$

so that the inequality (6.39) does not hold. Hence the radius of convergence is 1 and we end the proof of the problem. ∎

Problem 6.23

⋆ ⋆ *Let R be the radius of convergence of the power series $\displaystyle\sum_{n=0}^{\infty} c_n z^n$, where $c_n \neq 0$ for all $n = 1, 2, \ldots$. Prove that*

$$\liminf_{n\to\infty} \left| \frac{c_n}{c_{n+1}} \right| \le R \le \limsup_{n\to\infty} \left| \frac{c_n}{c_{n+1}} \right|. \quad (6.40)$$

Proof. If $0 < R < +\infty$, then we know from Theorem 6.20 that

$$\frac{1}{R} = \limsup_{n\to\infty} \sqrt[n]{|c_n|}. \quad (6.41)$$

Since $|c_n| > 0$ for all $n = 1, 2, \ldots$, it follows from Problem 5.13 that

$$\liminf_{n\to\infty} \left| \frac{c_{n+1}}{c_n} \right| \le \limsup_{n\to\infty} \sqrt[n]{|c_n|} \le \limsup_{n\to\infty} \left| \frac{c_{n+1}}{c_n} \right|. \quad (6.42)$$

It is obvious that

$$\frac{1}{\displaystyle\liminf_{n\to\infty} \left| \frac{c_{n+1}}{c_n} \right|} = \limsup_{n\to\infty} \left| \frac{c_n}{c_{n+1}} \right| \quad \text{and} \quad \frac{1}{\displaystyle\limsup_{n\to\infty} \left| \frac{c_{n+1}}{c_n} \right|} = \liminf_{n\to\infty} \left| \frac{c_n}{c_{n+1}} \right|, \quad (6.43)$$

Hence we obtain the desired result (6.40) by combining the inequality (6.42) and the expressions (6.41) and (6.43). This completes the proof of the problem. ∎

CHAPTER 7

Limits and Continuity of Functions

7.1 Fundamental Concepts

A real or complex function f is a function whose values $f(x)$ are real or complex respectively. The main references for this chapter are [3, §4.5 - §4.23, pp. 74 - 95], [5, Chap. 4 & 5, pp. 102 - 159], [6, Chap. 3], [13, Chap. 4, pp. 83 - 98] and [15, Chap. 9, pp. 211 - 250].

7.1.1 Limits of Functions

Suppose that X and Y are metric spaces with metrics d_X and d_Y respectively. Let $E \subseteq X$, $f : E \to Y$ be a function, p be a limit point of E and $q \in Y$. The notation[a]

$$\lim_{x \to p} f(x) = q \qquad (7.1)$$

means that *for every $\epsilon > 0$, there exists* a $\delta > 0$ such that

$$d_Y(f(x), q) < \epsilon$$

for all points $x \in E$ with

$$0 < d_X(x, p) < \delta. \qquad (7.2)$$

By the inequality (7.2), it is clear that $x \neq p$. Furthermore, we note that the number δ depends on ϵ. The notation (7.1) is read "the **limit of** $f(x)$ as x tends to p."

The following theorem relates limits of functions and limits of convergent sequences:

Theorem 7.1. *The limit (7.1) holds if and only if*

$$\lim_{n \to \infty} f(p_n) = q$$

for every sequence $\{p_n\} \subseteq E \setminus \{p\}$ such that $p_n \to p$.

---

[a]Sometimes we write $f(x) \to q$ as $x \to p$.

By Theorems 5.1 (Uniqueness of Limits of Sequences) and 7.1, we know that if f has a limit at the point p, then this limit must be **unique**.

Suppose that $f, g : X \to Y$ are two functions from X to Y. As usual, the **sum** $f + g$ is defined to be the function whose value at $x \in X$ is $f(x) + g(x)$. The **difference** $f - g$, the **product** fg and the **quotient** $\frac{f}{g}$ can be defined similarly. Similar to Theorem 5.5, we have the following algebra of limits of functions.

Theorem 7.2. *Suppose that X and Y are metric spaces, $E \subseteq X$ and p is a limit point of E. Let $f, g : E \to Y$,*

$$\lim_{x \to p} f(x) = A \quad and \quad \lim_{x \to p} g(x) = B.$$

Then we have

(a) $\displaystyle\lim_{x \to p} (f \pm g)(x) = A \pm B.$

(b) $\displaystyle\lim_{x \to p} (fg)(x) = AB.$

(c) $\displaystyle\lim_{x \to p} \left(\frac{f}{g}\right)(x) = \frac{A}{B}$ *provided that $B \neq 0$.*

In Chapter 5, we have the **Squeeze Theorem for Convergent Sequences** (Theorem 5.6). Similarly, we have a corresponding result for limits of functions (see [2, Theorem 3.3, p. 133]):

Theorem 7.3 (Squeeze Theorem for Limits of Functions). *Let $f, g, h : (a, b) \to \mathbb{R}$ and $p \in (a, b)$. Suppose that $g(x) \leq f(x) \leq h(x)$ for all $x \in (a, b) \setminus \{p\}$ and*

$$\lim_{x \to p} g(x) = \lim_{x \to p} h(x).$$

Then we have

$$\lim_{x \to p} f(x).$$

7.1.2 Continuity and Uniform Continuity of Functions

Definition 7.4 (Continuity). *Suppose that X and Y are metric spaces with metrics d_X and d_Y respectively. Let $p \in E \subseteq X$ and $f : E \to Y$. The function f is said to be **continuous at** p if for every $\epsilon > 0$, there exists a $\delta > 0$ such that*

$$d_Y(f(x), f(p)) < \epsilon$$

*for all $x \in E$ with $d_X(x, p) < \delta$. If f is continuous at every point of E, then f is said to be **continuous on** E.*

If p is a **limit point** of E, then Definition 7.4 (Continuity) implies that

$$\lim_{x \to p} f(x) = f(p).$$

In view of Theorem 7.1, the following result, which can be treated as an equivalent way of formulating the definition of continuity, is obvious:

Theorem 7.5. *Suppose that X and Y are metric spaces. Let $p \in E \subseteq X$ and $f : E \to Y$. The function f is **continuous at** p if and only if, for every sequence $\{x_n\} \subseteq E \setminus \{p\}$ converging to p, the corresponding sequence $\{f(x_n)\} \subseteq Y$ converging to $f(p)$, i.e.,*

$$\lim_{n\to\infty} f(x_n) = f\left(\lim_{n\to\infty} x_n \right).$$

> **Remark 7.1**
>
> It is well-known that every polynomial is continuous on its domain and a rational function is continuous at points wherever the denominator is nonzero. Furthermore, it follows from the triangle inequality that the function $f : E \subseteq \mathbb{R}^n \to \mathbb{R}$ defined by $f(\mathbf{x}) = |\mathbf{x}|$ is also continuous on E.

For compositions of functions, we see that a continuous function of a continuous function is always continuous. The detailed description is given in the following result:

Theorem 7.6. *Given that X, Y and Z are metric spaces. Let $E \subseteq X$. Suppose that $f : E \to Y$, $g : f(E) \to Z$ and $h = g \circ f : X \to Z$, i.e.,*

$$h(x) = g(f(x))$$

for every $x \in E$. If f is continuous at $p \in E$ and g is continuous at $f(p)$, then h is continuous at p.

Another useful characterization of the continuity of a function depends on the "topology" (open sets and closed sets) of the metric spaces X and Y. In fact, we have

Theorem 7.7. *Let $f : X \to Y$ be a function. Then f is continuous on X if and only if $f^{-1}(V)$ is open in X for every open set V in Y.*

> **Remark 7.2**
>
> The property "open set" in Theorem 7.7 can be replaced by the property "closed set".

Besides the continuity of functions, **uniform continuity of functions** is another important concept. The definition is as follows:

Definition 7.8 (Uniform Continuity). *Suppose that X and Y are metric spaces with metrics d_X and d_Y respectively. Then the function $f : X \to Y$ is said to be **uniformly continuous** on X if for every $\epsilon > 0$, there exists a $\delta > 0$ such that*

$$d_Y(f(x), f(y)) < \epsilon$$

for all $x, y \in X$ with $d_X(x, y) < \delta$.

If we compare Definitions 7.4 (Continuity) and 7.8 (Uniform Continuity), then we can see their differences easily. In fact, continuity of a function f is defined at a point $x \in X$, but uniform continuity of f is defined on the whole set X. In addition, the chosen δ in Definitions 7.4 (Continuity) depends on two factors: the point p and the initial value ϵ, while that in Definition 7.8 (Uniform Continuity) relies solely on the initial value ϵ and different points won't give different δ. **Obviously, a uniform continuous function f is continuous.**

▌ 7.1.3 The Extreme Value Theorem

There are many nice properties when the domain of a continuous function is a compact set. In fact, the next theorem shows that the image of a compact set under a continuous function f is always compact.

Theorem 7.9 (Continuity and Compactness). *Let $f : X \to Y$ be a continuous function from the metric space X to the metric space Y. If $E \subseteq X$ is compact, then $f(E) \subseteq Y$ is compact.*

We say that a function $\mathbf{f} : E \subseteq X \to \mathbb{R}^k$ is **bounded** on E if there exists a positive constant M such that
$$|\mathbf{f}(x)| \le M$$
for all $x \in E$. Thus, Theorem 7.9 (Continuity and Compactness) and the Heine-Borel Theorem imply that $\mathbf{f}(E)$ is bounded on E. In particular, when $k = 1$ (i.e., f is a real function), the Completeness Axiom implies that both
$$\sup f(E) \quad \text{and} \quad \inf f(E)$$
exist in $\mathbb{R}$.

The following famous result, namely the **Extreme Value Theorem**, shows that a continuous real function $f : E \to \mathbb{R}$ takes on the values $\sup f(E)$ and $\inf f(E)$.

The Extreme Value Theorem. *Suppose that $f : E \to \mathbb{R}$ is continuous, where E is compact. Let*
$$M = \sup_{x \in E} f(x) \quad \text{and} \quad m = \inf_{x \in E} f(x).$$
Then there exists $p, q \in E$ such that
$$f(p) = M \quad \text{and} \quad f(q) = m.$$

In §7.1.2, we know that a uniform continuous function f is continuous, but the converse is not always true. Interestingly, this is true when the domain is a compact set.

Theorem 7.10. *Suppose that X and Y are metric spaces and X is compact. If $f : X \to Y$ is continuous, then f is uniformly continuous on X.*

If f is bijective, then we can say more:

Theorem 7.11. *Suppose that X and Y are metric spaces and X is compact. If $f : X \to Y$ is bijective and continuous, then the **inverse function** $f^{-1} : Y \to X$ is continuous.*

▌ 7.1.4 The Intermediate Value Theorem

Similar to functions with compact domains, there are also many nice properties when the domain of the function f is connected. Actually, the following result says that continuity "preserves" connectedness.

Theorem 7.12 (Continuity and Connectedness). *Let $f : X \to Y$ be a continuous function from the metric space X to the metric space Y. If $E \subseteq X$ is connected, then $f(E) \subseteq Y$ is connected.*

Besides the **Extreme Value Theorem**, we have another important theorem for continuous functions. It is called the **Intermediate Value Theorem**.

The Intermediate Value Theorem. *Suppose that* $f : [a,b] \to \mathbb{R}$ *is continuous. If we have* $f(a) < c < f(b)$ *or* $f(b) < c < f(a)$, *then there exists a point* $x_0 \in (a,b)$ *at which* $f(x_0) = c$.

7.1.5 Discontinuity of Functions

Suppose that $f : X \to Y$ is a function. If f is *not* continuous at $p \in X$, then we say that f is **discontinuous** at p. In this case, classifications of all the discontinuous points of a function should be considered.

Definition 7.13 (Left-hand Limits and Right-hand Limits). *Suppose that* $f : (a,b) \to Y$ *and* $p \in [a,b)$, *where* Y *is a metric space. We define* $f(p+)$ *the **right-hand limit** of* f *at* p *by*

$$f(p+) = q$$

if $f(x_n) \to q$ *as* $n \to \infty$ *for all sequences* $\{x_n\} \subseteq (p,b)$ *converging to* p. *Similarly, if* $p \in (a,b]$, *we define* $f(p-)$ *the **left-hand limit** of* f *at* p *by*

$$f(p-) = q$$

if $f(x_n) \to q$ *as* $n \to \infty$ *for all sequences* $\{x_n\} \subseteq (a,p)$ *converging to* p.

Theorem 7.14. *Suppose that* $f : (a,b) \to Y$, *where* Y *is a metric space. Then* $\lim_{x \to p} f(x)$ *exists if and only if*

$$f(p+) = f(p-) = \lim_{x \to p} f(x).$$

In particular, f *is continuous at* $p \in (a,b)$ *if and only if*

$$f(p+) = f(p-) = f(p).$$

By this result, we can determine the type of the discontinuity (if it exists) of a mapping f as follows:

Definition 7.15 (Types of Discontinuity). *Suppose that* $f : (a,b) \to Y$, *where* Y *is a metric space. Suppose that* f *is discontinuous at* $p \in (a,b)$. *Then one of the following conditions is satisfied:*

(a) *Both* $f(p+)$ *and* $f(p-)$ *exist, but* $f(p+) = f(p-) \neq f(p)$.

(b) *Both* $f(p+)$ *and* $f(p-)$ *exist, but* $f(p+) \neq f(p-)$.

(c) *Either* $f(p+)$ *and* $f(p-)$ *does not exist.*

The function f is said to have a **discontinuity of the first kind** at p (or a **simple discontinuity**) in Case (a) or Case (b). For Case (c), f is said to have a **discontinuity of the second kind** at p.

Theorem 7.16 (Countability of Simple Discontinuities). *The set of all simple discontinuities of the function* $f : (a,b) \to \mathbb{R}$ *is **at most countable**.*

7.1.6 Monotonic Functions

There is a close connection between real monotonic functions $f : (a,b) \to \mathbb{R}$ and types of its discontinuities in the sense of Definition 7.15 (Types of Discontinuity).

Definition 7.17 (Monotonic Functions). *Suppose that $f : (a,b) \to \mathbb{R}$ is a function. We say f is **monotonically increasing** on (a,b) if*

$$f(x) \le f(y) \tag{7.3}$$

*for every $x < y$ with $x, y \in (a,b)$. Similarly, f is **monotonically decreasing** on (a,b) if*

$$f(x) \ge f(y) \tag{7.4}$$

for every $x < y$ with $x, y \in (a,b)$.[b]

Theorem 7.18. *Suppose that $f : (a,b) \to \mathbb{R}$ is monotonically increasing. Then both $f(p+)$ and $f(p-)$ exist for every $p \in (a,b)$ and we have*

$$f(p-) \le f(p) \le f(p+).$$

In other words, monotonic functions does not have discontinuities of the second kind.

> **Remark 7.3**
>
> Notice that analogous results of Theorem 7.18 hold for monotonically decreasing functions.

Theorem 7.19 (Froda's Theorem). *Suppose that the function $f : (a,b) \to \mathbb{R}$ is monotonic. Then the set of discontinuities is **at most countable**.*

7.1.7 Limits at Infinity

Sometimes we need to deal with limits at infinity, so we need the following definition:

Definition 7.20 (Limits at Infinity). *Suppose that $f : E \subseteq \mathbb{R} \to \mathbb{R}$. We say that $f(x)$ converges to L as $x \to +\infty$ in E if for every $\epsilon > 0$, there exists a constant M such that*

$$|f(x) - L| < \epsilon$$

for all $x \in E$ such that $x > M$. Analogous definition evidently holds when $x \to -\infty$.

By this definition, we can do the same things with limits at infinity as what we have done with limits at a point p. In short, it means that the analogue of Theorem 7.2 also holds if the operations of the numbers there are well-defined.

---

[b]If the inequality sign in (7.3) or (7.4) is strict, then we say f is **strictly increasing** or **strictly decreasing** on (a,b) respectively.

7.2 Limits of Functions

Problem 7.1

$(\star)$ Suppose that $f : (a, b) \to \mathbb{R}$ and $x \in \mathbb{R}$. If $\lim_{h \to 0} |f(x+h) - f(x)| = 0$, prove that

$$\lim_{h \to 0} |f(x+h) - f(x-h)| = 0.$$

Is the converse true?

Proof. By Theorem 7.1, we have

$$\lim_{n \to \infty} |f(x + h_n) - f(x)| = 0 \tag{7.5}$$

for every sequence $\{h_n\}$ such that $h_n \neq 0$ and $h_n \to 0$. By the triangle inequality, we have

$$|f(x + h_n) - f(x - h_n)| = |f(x + h_n) - f(x) + f(x) - f(x - h_n)|$$
$$\leq |f(x + h_n) - f(x)| + |f(x) - f(x - h_n)| \tag{7.6}$$

Apply Theorem 5.6 (Squeeze Theorem for Convergent Sequences) to the inequality (7.6) and then using the limit (7.5), we deduce

$$\lim_{n \to \infty} |f(x + h_n) - f(x - h_n)| = 0$$

for every sequence $\{h_n\}$ such that $h_n \neq 0$ and $h_n \to 0$. By Theorem 7.1, it means that

$$\lim_{h \to 0} |f(x + h) - f(x - h)| = 0$$

holds.

The converse is false. For example, consider $f : \mathbb{R} \to \mathbb{R}$ to be defined by

$$f(x) = \begin{cases} x^2, & \text{if } x \neq 0; \\ 1, & \text{if } x = 0. \end{cases}$$

Then we have

$$\lim_{h \to 0} |f(0 + h) - f(0 - h)| = \lim_{h \to 0} |f(h) - f(-h)| = 0$$

but

$$\lim_{h \to 0} |f(0 + h) - f(0)| = \lim_{h \to 0} |f(h) - 1| = 1.$$

This completes the proof of the problem. ∎

Problem 7.2

$(\star)$ Let $x \in \mathbb{R}$. Prove that

$$\lim_{n \to \infty} \left[\lim_{k \to \infty} \cos^{2k}(n! \pi x) \right] = \begin{cases} 1, & \text{if } x \in \mathbb{Q}; \\ 0, & \text{if } x \in \mathbb{R} \setminus \mathbb{Q}. \end{cases}$$

Proof. Suppose that $x \in \mathbb{R} \setminus \mathbb{Q}$. Then $n!x \notin \mathbb{Z}$ for any $n \in \mathbb{N}$ and thus $n!\pi x$ is not a multiple of π for every $n \in \mathbb{N}$. Therefore, we have

$$0 \le |\cos(n!\pi x)| < 1$$

which implies that

$$0 \le |\cos(n!\pi x)| < \delta \tag{7.7}$$

for some $\delta < 1$. Now we apply Theorems 5.6 (Squeeze Theorem for Convergent Sequences) and 5.7(c) to (7.7) to obtain

$$\lim_{k \to \infty} \cos^{2k}(n!\pi x) = 0$$

and hence

$$\lim_{n \to \infty} \left[\lim_{k \to \infty} \cos^{2k}(n!\pi x) \right] = 0.$$

Next, we suppose that $x \in \mathbb{Q}$. Then $x = \frac{p}{q}$ for some $p, q \in \mathbb{Z}$ and $q \ne 0$. Take $n > q$ so that $n!x \in \mathbb{N}$ implying that

$$|\cos(n!\pi x)| = 1. \tag{7.8}$$

Now the expression (7.8) shows easily that

$$\lim_{k \to \infty} \cos^{2k}(n!\pi x) = 1$$

and hence

$$\lim_{n \to \infty} \left[\lim_{k \to \infty} \cos^{2k}(n!\pi x) \right] = 1.$$

This ends the proof of the problem. ■

> **Remark 7.4**
>
> The limit in Problem 7.2 is the analytical form of the so-called the **Dirichlet function**.

Problem 7.3

(★) Suppose that $a_0, a_1, \ldots, a_n, b_0, b_1, \ldots, b_m \in \mathbb{R}$ and $a_0 b_0 \ne 0$. Let

$$R(x) = \frac{a_0 x^n + a_1 x^{n-1} + \cdots + a_n}{b_0 x^m + b_1 x^{m-1} + \cdots + b_m}.$$

Prove that

$$\lim_{x \to \infty} R(x) = \begin{cases} +\infty, & \text{if } n > m; \\[2mm] \dfrac{a_0}{b_0}, & \text{if } n = m; \\[2mm] 0, & \text{if } n < m. \end{cases}$$

Proof. Express $R(x)$ in the following form

$$R(x) = \frac{a_0 x^{n-m} + a_1 x^{n-m-1} + \cdots + a_n x^{-m}}{b_0 + b_1 x^{-1} + \cdots + b_m x^{-m}}. \tag{7.9}$$

When $n > m$, the denominator and the numerator of (7.9) tend to b_0 and ∞ respectively as $x \to \infty$. Therefore, we have

$$\lim_{x \to \infty} R(x) = \infty.$$

When $n = m$, the term $a_0 x^{n-m}$ becomes a_0 and the other terms in the numerator still have the factor x^{-1}. Thus the denominator and the numerator of (7.9) tend to b_0 and a_0 respectively as $x \to \infty$. Hence we have

$$\lim_{x \to \infty} R(x) = \frac{a_0}{b_0}$$

in this case.

When $n < m$, since each term in the numerator involves the factor x^{-1}, each term tends to 0 as $x \to \infty$. In other words, we have

$$\lim_{x \to \infty} R(x) = 0$$

in this case, completing the proof of the problem. ∎

Problem 7.4

(⋆) *Let $n \in \mathbb{N}$ and $a \in \mathbb{R}$, evaluate the limit*

$$\lim_{x \to a} \frac{(x^n - a^n) - na^{n-1}(x - a)}{(x - a)^2}. \tag{7.10}$$

Proof. Put $x = a + y$. Now $x \to a$ if and only if $y \to 0$. By the Binomial Theorem, the limit (7.10) becomes

$$\lim_{y \to 0} \frac{(a+y)^n - a^n - na^{n-1}y}{y^2} = \lim_{y \to 0} \left(\frac{a^n + C_1^n a^{n-1}y + C_2^n a^{n-2}y^2 + \cdots + y^n - a^n - na^{n-1}y}{y^2} \right)$$

$$= \lim_{y \to 0}(C_2^n a^{n-2} + C_3^n a^{n-3}y + \cdots + y^{n-2})$$

$$= C_2^n a^{n-2}.$$

This finishes the proof of the problem. ∎

Problem 7.5

(⋆)(⋆) *Suppose that $\lim_{x \to 0} f(x) = A$ and $\lim_{x \to A} g(x) = B$. Prove or disprove*

$$\lim_{x \to 0} g(f(x)) = B.$$

Proof. Consider the following functions[c]

$$f(x) = \begin{cases} \frac{1}{q}, & \text{if } x = \frac{p}{q}, p \in \mathbb{Z}, q \in \mathbb{N} \text{ and } p, q \text{ are coprime}; \\ \\ 0, & x \in \mathbb{R} \setminus \mathbb{Q} \end{cases}$$

---

[c]The function is called the **Riemann function** and some properties of this function will be given in Problem 9.8.

and

$$g(x) = \begin{cases} 1, & \text{if } x \neq 0; \\ -1, & \text{if } x = 0. \end{cases}$$

Now it is easy to check that

$$\lim_{x \to 0} f(x) = 0 \quad \text{and} \quad \lim_{x \to 0} g(x) = 1.$$

However, we claim that the limit

$$\lim_{x \to 0} g(f(x)) \tag{7.11}$$

does not exist. To see this, if $\{x_n\} \subseteq \mathbb{Q} \setminus \{0\}$ and $x_n \to 0$, then we have $f(x_n) \neq 0$ for every $n \in \mathbb{N}$ and this implies that

$$g(f(x_n)) = 1 \tag{7.12}$$

for every $n \in \mathbb{N}$. Next, if we choose $\{y_n\} \subseteq \mathbb{R} \setminus (\mathbb{Q} \cup \{0\})$ and $y_n \to 0$, then we have $f(y_n) = 0$ for every $n \in \mathbb{N}$. In this case, we have

$$g(f(x_n)) = -1 \tag{7.13}$$

for every $n \in \mathbb{N}$. Since the two values (7.12) and (7.13) are not equal, it follows from Theorem 7.1 that the limit (7.11) does not exist. This completes the proof of the problem. ∎

Problem 7.6

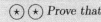 *Prove that*

$$\lim_{x \to \infty} \left(1 + \frac{1}{x}\right)^x = e.$$

Proof. Fix $x \in [1, \infty)$. Let n_x be the *unique* integer such that

$$n_x \leq x < n_x + 1.$$

Now we have

$$\left(1 + \frac{1}{x}\right)^x \leq \left(1 + \frac{1}{n_x}\right)^{n_x + 1} = \left(1 + \frac{1}{n_x}\right)^{n_x} \left(1 + \frac{1}{n_x}\right) \tag{7.14}$$

and

$$\left(1 + \frac{1}{x}\right)^x \geq \left(1 + \frac{1}{n_x + 1}\right)^{n_x} = \left(1 + \frac{1}{n_x + 1}\right)^{n_x + 1} \left(1 + \frac{1}{n_x + 1}\right)^{-1}. \tag{7.15}$$

Combining the inequalities (7.14) and (7.15), we get

$$\left(1 + \frac{1}{n_x + 1}\right)^{n_x + 1} \left(1 + \frac{1}{n_x + 1}\right)^{-1} \leq \left(1 + \frac{1}{x}\right)^x \leq \left(1 + \frac{1}{n_x}\right)^{n_x} \left(1 + \frac{1}{n_x}\right). \tag{7.16}$$

It is clear that $n_x \to \infty$ if and only if $x \to \infty$ so that

$$\lim_{n_x \to \infty} \left(1 + \frac{1}{x}\right)^x = \lim_{x \to \infty} \left(1 + \frac{1}{x}\right)^x.$$

Furthermore, since

$$\lim_{n_x \to \infty} \left(1 + \frac{1}{n_x}\right) = \lim_{n_x \to \infty} \left(1 + \frac{1}{n_x + 1}\right)^{-1} = 1 \tag{7.17}$$

and[d]

$$\lim_{n_x \to \infty} \left(1 + \frac{1}{n_x}\right)^{n_x} = \lim_{n_x \to \infty} \left(1 + \frac{1}{n_x + 1}\right)^{n_x+1} = \mathrm{e}, \tag{7.18}$$

we apply the results (7.17), (7.18) and Theorem 5.6 (Squeeze Theorem for Convergent Sequences) to the inequalities (7.16) to obtain the desired result

$$\lim_{x \to \infty} \left(1 + \frac{1}{x}\right)^{x} = \lim_{n_x \to \infty} \left(1 + \frac{1}{n_x}\right)^{n_x} = \mathrm{e}.$$

This finishes the proof of the problem. ∎

Problem 7.7

$\bigstar$ $\bigstar$ *Suppose that* $f : (0, 1) \to \mathbb{R}$ *satisfies the conditions*

$$\lim_{x \to 0} f(x) = 0 \quad and \quad \lim_{x \to 0} \frac{f(2x) - f(x)}{x} = 0.$$

Prove that

$$\lim_{x \to 0} \frac{f(x)}{x} = 0.$$

Proof. Given that $\epsilon > 0$. By the second condition, there exists a $\delta > 0$ such that

$$\left|\frac{f(2x) - f(x)}{x}\right| < \epsilon \tag{7.19}$$

for all $x \in (0, \delta)$. It is easy to see that $\frac{x}{2^n} \in (0, \delta)$ for every positive integer n. Therefore, it follows from the inequality (7.19) that

$$\left|f\left(\frac{x}{2^{n-1}}\right) - f\left(\frac{x}{2^n}\right)\right| < \epsilon \times \frac{x}{2^n} \tag{7.20}$$

for every positive integer n. By the triangle inequality and the inequality (7.20), we follow that

$$\left|f(x) - f\left(\frac{x}{2^n}\right)\right| \leq \underbrace{\left|f(x) - f\left(\frac{x}{2}\right)\right| + \left|f\left(\frac{x}{2}\right) - f\left(\frac{x}{2^2}\right)\right| + \cdots + \left|f\left(\frac{x}{2^{n-1}}\right) - f\left(\frac{x}{2^n}\right)\right|}_{n \text{ terms}}$$

$$< \epsilon x \left(\frac{1}{2} + \frac{1}{2^2} + \cdots + \frac{1}{2^n}\right)$$
$$< \frac{\epsilon x}{2} \left(1 + \frac{1}{2} + \frac{1}{2^2} + \cdots\right)$$
$$= \epsilon x \tag{7.21}$$

for all $x \in (0, \delta)$. By the first condition and Theorem 7.1, we have

$$\lim_{n \to \infty} \left|f\left(\frac{x}{2^n}\right)\right| = \lim_{x \to 0} |f(x)| = 0. \tag{7.22}$$

By the triangle inequality and the inequality (7.21), we know that

$$|f(x)| \leq \left|f(x) - f\left(\frac{x}{2^n}\right)\right| + \left|f\left(\frac{x}{2^n}\right)\right| < \epsilon x + \left|f\left(\frac{x}{2^n}\right)\right| \tag{7.23}$$

---

[d]See the proof of Problem 5.14.

for all $x \in (0, \delta)$. By taking $n \to \infty$ to both sides of the inequality (7.23) and then applying the result (7.22), we gain

$$|f(x)| \leq \epsilon x$$

or equivalently,

$$\left| \frac{f(x)}{x} \right| \leq \epsilon$$

for all $x \in (0, \delta)$. By definition, it means that

$$\lim_{x \to 0} \frac{f(x)}{x} = 0,$$

completing the proof of the problem.   ∎

7.3 Continuity and Uniform Continuity of Functions

Problem 7.8

⍟⍟ *Suppose that $f : [0, 1] \to \mathbb{R}$ is continuous and $f(x) = 0$ for every irrational x in $[0, 1]$. Prove that $f \equiv 0$ on $[0, 1]$.*

Proof. By Problem 2.9 and Remark 2.1, we know that given any $x \in [0, 1]$, there exists a sequence $\{\theta_n\} \subseteq [0, 1]$ such that

$$\lim_{n \to \infty} \theta_n = x.$$

Since f is continuous on $[0, 1]$, it follows from Theorem 7.5 that

$$f(x) = f\left(\lim_{n \to \infty} \theta_n \right) = \lim_{n \to \infty} f(\theta_n) = 0.$$

Hence we have $f \equiv 0$ on $[0, 1]$. This completes the proof of the problem.   ∎

Problem 7.9

⍟⍟ *Suppose that $f : \mathbb{R} \to \mathbb{R}$ and f is continuous at 0. If $f(x + y) = f(x) + f(y)$ for every $x, y \in \mathbb{R}$, prove that there exists a constant a such that*

$$f(x) = ax$$

for all $x \in \mathbb{R}$.

Proof. Since f is continuous at 0, for every $\epsilon > 0$, there exists a $\delta > 0$ such that

$$|f(h) - f(0)| < \epsilon \tag{7.24}$$

for all $|h| < \delta$. By the hypothesis and the inequality (7.24), we have

$$|f(h)| = |[f(h) + f(0)] - f(0)| = |f(h + 0) - f(0)| < \epsilon \tag{7.25}$$

for all $|h| < \delta$. Let $p \in \mathbb{R}$. Then it follows from the hypothesis and the inequality (7.25) that

$$|f(p + h) - f(p)| = |f(p) + f(h) - f(p)| = |f(h)| < \epsilon$$

for all $|h| < \delta$. By Definition 7.4 (Continuity), the function f is continuous at every point $p \in \mathbb{R}$.

Let $a = f(1)$. Then it can be proven by induction that

$$f(n) = na \tag{7.26}$$

for every positive integer n. Since $f(0) = f(0 + 0) = f(0) + f(0)$, we have

$$f(0) = 0. \tag{7.27}$$

Furthermore, the hypothesis and the fact (7.27) imply that

$$f(-x) + f(x) = f(-x + x) = f(0) = 0$$

so that

$$f(-x) = -f(x) \tag{7.28}$$

for every $x \in \mathbb{R}$.

Now, by combining the facts (7.26), (7.27) and (7.28), we deduce that the formula (7.26) holds for every integer n. If $r = \frac{p}{q}$, where $p \in \mathbb{Z}$ and $q \in \mathbb{N}$, then we see from the hypothesis and the expression (7.26) that

$$\underbrace{f\left(\frac{p}{q}\right) + \cdots + f\left(\frac{p}{q}\right)}_{q \text{ terms}} = f\left(q \cdot \frac{p}{q}\right) = f(p) = ap$$

which implies that

$$f\left(\frac{p}{q}\right) = a \times \frac{p}{q},$$

i.e., $f(r) = ar$ for every rational r. Finally, by an argument similar to the proof of Problem 7.8, we can show that

$$f(x) = ax$$

for all $x \in \mathbb{R}$. Hence we have finished the proof of the problem. ■

Problem 7.10

⋆ ⋆ *Find all continuous functions $f : \mathbb{R} \to \mathbb{R}$ such that*

$$f(x) + f(3x) = 0 \tag{7.29}$$

holds for all $x \in \mathbb{R}$.

Proof. The hypothesis implies that $f(0) = 0$. Furthermore, it is clear that $f \equiv 0$ satisfies the hypothesis (7.29). We claim that $f \equiv 0$ is the *only* function satisfying the requirement (7.29).

To this end, we prove that

$$f(x) = (-1)^n f\left(\frac{x}{3^n}\right) \tag{7.30}$$

for every positive integer n and $x \in \mathbb{R}$. We use induction. The case $n = 1$ follows from the hypothesis directly. Assume that

$$f(x) = (-1)^k f\left(\frac{x}{3^k}\right) \tag{7.31}$$

for some positive integer k and all $x \in \mathbb{R}$. If $n = k + 1$, then we replace x by $\frac{x}{3^{k+1}}$ in the hypothesis (7.29) and we obtain

$$f\left(\frac{x}{3^{k+1}}\right) + f\left(\frac{x}{3^k}\right) = 0. \tag{7.32}$$

By putting the assumption (7.31) into the expression (7.32), we see that for all $x \in \mathbb{R}$,

$$f\left(\frac{x}{3^{k+1}}\right) + (-1)^k f(x) = 0$$

$$f(x) = (-1)^{k+1} f\left(\frac{x}{3^{k+1}}\right).$$

Thus the statement (7.30) is true for $n = k + 1$ if it is true for $n = k$. By induction, we are able to prove that the formula (7.30) is true for all $n \in \mathbb{N}$ and $x \in \mathbb{R}$.

Since f is continuous at 0, we deduce from Theorem 7.5 that

$$\lim_{n \to \infty} f\left(\frac{x}{3^n}\right) = f(0) = 0 \tag{7.33}$$

for every $x \in \mathbb{R}$. Combining the formula (7.30) and the result (7.33), we derive that

$$f(x) = 0$$

for all $x \in \mathbb{R}$, i.e., $f \equiv 0$ which is the *only* function satisfying the hypothesis (7.29). We finish the proof of the problem. ∎

Problem 7.11

⋆ Prove that the Dirichlet function $D(x)$ in Problem 7.2 is discontinuous on $\mathbb{R}$.

Proof. By the definition in Problem 7.2, we see that

$$D(x) = \begin{cases} 1, & \text{if } x \in \mathbb{Q}; \\ 0, & \text{if } x \in \mathbb{R} \setminus \mathbb{Q}. \end{cases}$$

Let $x_0 \in \mathbb{R}$. Then *for every* $\delta > 0$, the interval $(x_0 - \delta, x_0 + \delta)$ contains a rational number q and an irrational number θ.[e] Thus if $x_0 \in \mathbb{Q}$, then $D(x_0) = 1$ and we may take an irrational $\theta \in (x_0 - \delta, x_0 + \delta)$ so that

$$|D(x_0) - D(\theta)| = |1 - 0| = 1.$$

---

[e]See Theorem 2.2 (Density of Rationals) and Problem 2.9.

Otherwise, if $x_0 \in \mathbb{R} \setminus \mathbb{Q}$, then $D(x_0) = 0$ and we may take a rational $q \in (x_0 - \delta, x_0 + \delta)$ so that

$$|D(x_0) - D(q)| = |0 - 1| = 1.$$

In other words, the function $D(x)$ does not satisfy Definition 7.4 (Continuity). Therefore, $D(x)$ is discontinuous at x_0 and hence it is discontinuous on $\mathbb{R}$. We complete the proof of the problem. ∎

Problem 7.12

(⋆) Suppose that X and Y are metric spaces. Let $f : X \to Y$ and $g : X \to Y$ be continuous on X. Define $\varphi(x) : X \to Y$ and $\psi(x) : X \to Y$ by

$$\varphi(x) = \max(f(x), g(x)) \quad \text{and} \quad \psi(x) = \min(f(x), g(x))$$

for every $x \in X$ respectively. Prove that φ and ψ are continuous on X.

Proof. By the following identities

$$\varphi(x) = \max(f(x), g(x)) = \frac{1}{2}[f(x) + g(x) + |f(x) - g(x)|]$$

and

$$\psi(x) = \min(f(x), g(x)) = \frac{1}{2}[f(x) + g(x) - |f(x) - g(x)|],$$

we see from Theorem 7.2 and Remark 7.1 that φ and ψ are continuous at every point $x \in X$. Hence we have completed the proof of the problem. ∎

Problem 7.13

(⋆) Suppose that $f : X \to \mathbb{R}$ is continuous and $a \in \mathbb{R}$, where X is a metric space. Define $f_a : X \to \mathbb{R}$ by

$$f_a(x) = \begin{cases} a, & \text{if } f(x) > a; \\ f(x), & \text{if } |f(x)| \le a; \\ -a, & \text{if } f(x) < -a, \end{cases}$$

where $x \in X$. Prove that f_a is continuous on X.

Proof. It is clear from the definition that

$$f_a(x) = \max(\min(a, f(x)), -a).$$

Hence we deduce from Problem 7.12 that f_a is continuous on X. ∎

Problem 7.14

$(\star)$ *Let X be a metric space with metric d_X and $f : X \to X$ be continuous. Suppose that*

$$d_X(f(x), f(y)) \geq d_X(x, y) \tag{7.34}$$

for all $x, y \in X$. Prove that f is one-to-one and $f^{-1} : f(X) \to X$ is continuous.

Proof. If $x_1 \neq x_2$, then the inequality (7.34) implies that

$$d_X(f(x_1), f(x_2)) \geq d_X(x_1, x_2) > 0.$$

By definition (§1.1.2), we see that f is one-to-one. Recall that $f^{-1}(f(x)) = x$ for every $x \in X$. It follows from this and the inequality (7.34) that

$$d_X(f^{-1}(p), f^{-1}(q)) \leq d_X(f(f^{-1}(p)), f(f^{-1}(q))) = d_X(p, q)$$

for every $p, q \in f(X)$. Thus we conclude from Definition 7.4 (Continuity) that f^{-1} is continuous on $f(X)$, completing the proof of the problem. ∎

Problem 7.15 (The Sign-preserving Property)

$(\star)(\star)$ *Suppose that $f : E \subseteq \mathbb{R} \to \mathbb{R}$ is continuous at p and $f(p) \neq 0$. Then there exists a $\delta > 0$ such that f has the same sign as $f(p)$ in $(p - \delta, p + \delta)$.*

Proof. By the continuity of f at p, for every $\epsilon > 0$, there exists a $\delta > 0$ such that

$$f(p) - \epsilon < f(x) < f(p) + \epsilon \tag{7.35}$$

whenever $x \in (p - \delta, p + \delta)$. Put $\epsilon = \frac{f(p)}{2}$ into the inequalities (7.35), we see that

$$\frac{1}{2}f(p) < f(x) < \frac{3}{2}f(p)$$

on $(p - \delta, p + \delta)$. Hence this shows that f has the same sign as $f(p)$ in $(p - \delta, p + \delta)$, completing the proof of the problem. ∎

Problem 7.16

$(\star)$ *If $f(x) = x^4$ on $\mathbb{R}$, prove that f is not uniformly continuous on $\mathbb{R}$.*

Proof. Fix $\epsilon = 1$. For every $\delta > 0$, Theorem 2.1 (The Archimedean Property) implies that there exists a positive integer N such that

$$N\delta^4 > 1. \tag{7.36}$$

Take $x = N\delta$ and $(N + \frac{1}{2})\delta$. Then it is easy to see that

$$|x - y| = \left| N\delta - \left(N + \frac{1}{2} \right)\delta \right| < \delta.$$

Now we have from the inequality (7.36) that

$$
\begin{aligned}
|f(x) - f(y)| &= |x^4 - y^4| \\
&= |x - y||x + y||x^2 + y^2| \\
&= \frac{\delta}{2} \times \left(2N + \frac{1}{2}\right)\delta \times \left[N^2 + \left(N + \frac{1}{2}\right)^2\right]\delta^2 \\
&> \frac{1}{4}(4N + 1)\delta^4 \\
&> N\delta^4 \\
&> 1.
\end{aligned}
$$

Hence we follow from Definition 7.8 (Uniform Continuity) that f is not uniformly continuous. This ends the proof of the problem. ∎

Problem 7.17

$(\star)$ *Construct a continuous real function f defined on $(0,1)$ but not uniformly continuous.*

Proof. The function $f : (0,1) \to \mathbb{R}$ defined by

$$
f(x) = \frac{1}{x}
$$

is obviously continuous on $(0,1)$ because it is the quotient of two polynomials and the denominator does not vanish in $(0,1)$. See Remark 7.1.

Assume that f was uniformly continuous on $(0,1)$. We pick $\epsilon = 1$. *For every* $\delta > 0$, if we take $x = \min(\delta, 1)$ and $y = \frac{x}{2}$, then we always have $x, y \in (0,1)$ and

$$
|x - y| = \frac{x}{2} < \delta,
$$

but

$$
\left|\frac{1}{x} - \frac{1}{y}\right| = \left|\frac{1}{x} - \frac{2}{x}\right| = \frac{1}{x} \geq 1 = \epsilon.
$$

Hence it contradicts Definition 7.8 (Uniform Continuity) and so f is not uniformly continuous on $(0,1)$, finishing the proof of the problem. ∎

Remark 7.5

Problems 7.16 and 7.17 tell us that the condition "compact" in Theorem 7.10 cannot be dropped.

Problem 7.18

$(\star)(\star)$ *Let $a > 0$. If $f : [a, +\infty) \to \mathbb{R}$ is continuous and*

$$
\lim_{x \to +\infty} f(x)
$$

exists, prove that f is uniformly continuous.

Proof. Given that $\epsilon > 0$. Since $\lim\limits_{x \to +\infty} f(x)$ exists, there exists a $M > a$ such that $x, y > M$ implies[f]

$$|f(x) - f(y)| < \epsilon.$$

Since f is continuous on $[a, M+1]$, it is uniformly continuous there. Thus there exists a $\delta > 0$ such that $x, y \in [a, M+1]$ with $|x - y| < \delta$ implies

$$|f(x) - f(y)| < \epsilon.$$

Let $\delta' = \min(\delta, 1)$ and $x, y \in [a, +\infty)$ with $|x - y| < \delta' \le \delta$. If $x \in [a, M]$ and $y \in (M+1, +\infty)$, then it is evident that

$$|x - y| \ge |y| - |x| > M + 1 - M = 1,$$

a contradiction. Thus if $x \in [a, M]$, then y is forced to lie in $[a, M+1]$. In this case, we have

$$x, y \in [a, M+1]. \tag{7.37}$$

Similarly, if $y \in (M+1, \infty)$, then x is forced to lie in $(M+1, +\infty)$ too. Therefore, we have the case

$$x, y \in (M+1, +\infty) \subset (M, +\infty). \tag{7.38}$$

Hence we deduce from the set relations (7.37) and (7.38) that

$$|f(x) - f(y)| < \epsilon.$$

By Definition 7.8 (Uniform Continuity), f is uniformly continuous on $[a, +\infty)$, completing the proof of the problem. ∎

Problem 7.19

$\bigstar$ $\bigstar$ Let $f : (a, b) \to \mathbb{R}$ be uniformly continuous. Prove that f is bounded on (a, b).

Proof. Take $\epsilon = 1$. Since f is uniformly continuous on (a, b), there is a $\delta > 0$ such that

$$|f(x) - f(y)| < 1$$

for all $x, y \in (a, b)$ with $|x - y| < \delta$. Fix this δ.

We note that *for every* $x \in (a, b)$, we have

$$\left| x - \frac{1}{2}(a + b) \right| < \frac{1}{2}(b - a) = \left(\frac{b - a}{2\delta} \right) \times \delta. \tag{7.39}$$

By Theorem 2.1 (The Archimedean Property), we can find a positive integer N such that

$$N > \frac{b - a}{2\delta}. \tag{7.40}$$

We *fix* this N. Combining the inequalities (7.39) and (7.40), we know that

$$\left| x - \frac{1}{2}(a + b) \right| < N\delta \tag{7.41}$$

---

[f]See Definition 7.20.

for every $x \in (a, b)$. Geometrically, the inequality (7.41) means that the distance between the point x and the mid-point of the interval (a, b) is always less than N times of the *fixed* length δ.

By the inequality (7.41), we can divide the interval with endpoints x and $\frac{1}{2}(a + b)$ into n subintervals each of length less than δ. It is clear that $n \leq N$. More precisely, for every $x \in (a, b)$, we can construct a sequence $\{x_1, x_2, \ldots, x_n\} \subseteq (a, b)$ of $n(\leq N)$ points such that

$$x_1 = \frac{1}{2}(a + b), \ldots, x_n = x \quad \text{or} \quad x_1 = x, \ldots, x_n = \frac{1}{2}(a + b),$$

where $|x_k - x_{k+1}| < \delta$ for $k = 1, 2, \ldots, n - 1$. This construction and the triangle inequality give

$$\left| f(x) - f\left(\frac{1}{2}(a + b)\right) \right| = |f(x_n) - f(x_1)|$$
$$\leq |f(x_n) - f(x_{n-1})| + |f(x_{n-1}) - f(x_{n-2})| + \cdots + |f(x_2) - f(x_1)|$$
$$< \underbrace{1 + 1 + \cdots + 1}_{n \text{ terms}}$$
$$= n \tag{7.42}$$

for every $x \in (a, b)$. Apply the triangle inequality again, it follows from the inequality (7.42) that

$$|f(x)| \leq \left| f(x) - f\left(\frac{1}{2}(a + b)\right) \right| + \left| f\left(\frac{1}{2}(a + b)\right) \right| < n + f\left(\frac{1}{2}(a + b)\right)$$

holds for every $x \in (a, b)$. This completes the proof of the problem. ∎

Problem 7.20

⊛ ⊛ Suppose that $f : E \subseteq \mathbb{R} \to \mathbb{R}$ is uniformly continuous on E. If $\{p_n\}$ is a Cauchy sequence in E, show that $\{f(p_n)\}$ is a Cauchy sequence in $\mathbb{R}$. Can the condition "uniformly continuous" be replaced by "continuous"?

Proof. By Definition 7.8 (Uniform Continuity), given $\epsilon > 0$, there exists a $\delta > 0$ such that

$$|f(x) - f(y)| < \epsilon \tag{7.43}$$

for all $x, y \in E$ with $|x - y| < \delta$. Since $\{p_n\}$ is Cauchy, there exists a positive integer N such that $m, n \geq N$ implies that

$$|p_m - p_n| < \delta. \tag{7.44}$$

Therefore, for $m, n \geq N$, we establish from the inequalities (7.43) and (7.44) that

$$|f(p_m) - f(p_n)| < \epsilon.$$

By Definition 5.12, $\{f(p_n)\}$ is also Cauchy.

We note that the conclusion does not hold anymore if the condition "uniformly continuous" is replaced by "continuous". For example, let $f : (0, 1) \to \mathbb{R}$ be given by

$$f(x) = \frac{1}{x}$$

for $x \in (0, 1)$. Then f is clearly continuous on $(0, 1)$ and the sequence $\{\frac{1}{n}\} \subseteq (0, 1)$ is Cauchy by Problem 5.19. However, $\{f(\frac{1}{n})\} = \{n\}$ which is a divergent sequence. Hence we have finished the proof of the problem. ∎

Problem 7.21

$\bigstar$ $\bigstar$ Suppose that X and Y are metric spaces and $f : X \to Y$. Prove that f is continuous on X if and only if

$$f^{-1}(V^\circ) \subseteq [f^{-1}(V)]^\circ$$

for every $V \subseteq Y$.

Proof. Suppose that f is continuous on X. Let $V \subseteq Y$. Since V° is open in Y, Theorem 7.7 tells us that $f^{-1}(V^\circ)$ is open in X. Since $f^{-1}(V^\circ) \subseteq f^{-1}(V)$ and we recall from Problem 4.11(c) $[f^{-1}(V)]^\circ$ is the largest open subset of $f^{-1}(V)$, we must have

$$f^{-1}(V^\circ) \subseteq [f^{-1}(V)]^\circ. \tag{7.45}$$

Conversely, we suppose that the set relation (7.45) holds for every $V \subseteq Y$. Let U be an open set in Y. Then Problem 4.11(b) implies that $U^\circ = U$ and so we follow from the set relation (7.45) that

$$f^{-1}(U) = f^{-1}(U^\circ) \subseteq [f^{-1}(U)]^\circ.$$

Since $[f^{-1}(U)]^\circ \subseteq f^{-1}(U)$, we see that

$$f^{-1}(U) = [f^{-1}(U)]^\circ$$

and it deduces from Problem 4.11(b) that $f^{-1}(U)$ is open in X. By Theorem 7.7, f is continuous on X, finishing the proof of the problem. ∎

7.4 The Extreme Value Theorem and the Intermediate Value Theorem

Problem 7.22

$\bigstar$ $\bigstar$ A function f is called **periodic** if, for some nonzero constant T, we have

$$f(x + T) = f(x)$$

for all values of x in the domain of f. Suppose that $f : \mathbb{R} \to \mathbb{R}$ is a nonempty, continuous and periodic function. Prove that f attains its supremum and infimum.

Proof. Let $T > 0$ be a period of f. Since f is continuous on $\mathbb{R}$, it is also continuous on $I = [0, T]$. By the Extreme Value Theorem, there exists $p, q \in I$ such that

$$f(p) = \sup_{x \in I} f(x) \quad \text{and} \quad f(q) = \inf_{x \in I} f(x). \tag{7.46}$$

For any $y \in \mathbb{R}$, we know that $y = x + nT$ for some integer n and $x \in I$, so the periodicity of f implies that

$$f(y) = f(x + nT) = f(x), \tag{7.47}$$

where $y \in \mathbb{R}$ and $x \in I$. Thus we follow from the results (7.46) and (7.47) that

$$\sup_{y \in \mathbb{R}} f(y) = \sup_{x \in I} f(x) = f(p) \quad \text{and} \quad \inf_{y \in \mathbb{R}} f(y) = \inf_{x \in I} f(x) = f(q).$$

This ends the proof of the problem. ∎

Problem 7.23

⋆ ⋆ A function $f : [a,b] \to \mathbb{R}$ is said to have a **local maximum** at p if there exists a neighborhood $N_\delta(p)$ of p such that $f(x) \le f(p)$ for all $x \in N_\delta(p) \cap [a,b]$. The concept of **local minimum** at p can be defined similarly.

Suppose that $f : [a,b] \to \mathbb{R}$ is continuous and has a local minimum at p and at q, where $p < q$. Prove that f has a local maximum at r, where $r \in (p,q)$.

Proof. Let $E = [p,q] \subseteq [a,b]$. Since f is continuous on $[a,b]$, it is also continuous on E. By the Extreme Value Theorem, there exists a $r \in E$ such that

$$f(r) = \sup_{x \in E} f(x). \tag{7.48}$$

If $p < r < q$, then since the supremum (7.48) indicates that $f(r)$ is the maximum value of f on E, it is definitely a local maximum of f. If $r = p$, then since $f(p)$ is a local minimum, there exists a $\delta > 0$ such that $p < x < p + \delta < q$ implies

$$f(p) \le f(x) \le f(r),$$

i.e., $f(x) = f(p) = f(r)$ for all $x \in (p, p+\delta) \subseteq (p,q)$. Now we may pick any $r' \in (p, p+\delta) \subseteq (p,q)$ so that

$$f(r') = f(r) = \sup_{x \in E} f(x),$$

i.e., f has a local maximum at $r' \in (p,q)$. The case for $r = q$ is similar, so we omit the details here. Thus we have completed the proof of the problem. ∎

Problem 7.24

⋆ ⋆ ⋆ Prove, without using Theorem 7.10, if $f : [a,b] \to \mathbb{R}$ is continuous, then f is uniformly continuous on $[a,b]$.

Proof. The following argument is due to Lüroth [9]. Given that $\epsilon > 0$. For each $p \in [a,b]$, define

$$E_p = \{\delta > 0 \,|\, |f(x) - f(y)| < \epsilon \text{ for all } x, y \in [p - \tfrac{\delta}{2}, p + \tfrac{\delta}{2}] \subseteq [a,b]\}$$

and

$$\delta(p) = \sup E_p.$$

In other words, $\delta(p)$ is the length of the *largest* interval $I(p)$ centred at p and

$$|f(x) - f(y)| < \epsilon$$

for all $x, y \in I(p)$. Since f is continuous at p, Definition 7.4 (Continuity) shows that $E_p \ne \varnothing$ for every $p \in [a,b]$. There are two cases for consideration:

- **Case (1):** $\delta(p) = \infty$ **for some** $p \in [a, b]$. Then any $\delta \in E_p$ satisfies Definition 7.8 (Uniform Continuity). We are done in this case.

- **Case (2):** $\delta(p) < \infty$ **for all** $p \in [a, b]$. We claim that $\delta(p)$ is continuous. To this end, consider the intervals $I(p)$ and $I(p + \omega)$ centred at points p and $p + \omega$, where $\omega > 0$. If $p + \omega - \frac{\delta(p+\omega)}{2} \leq p - \frac{\delta(p)}{2}$, then we have

$$p + \frac{\delta(p)}{2} \leq p + \omega + \frac{\delta(p)}{2} \leq p + \frac{\delta(p + \omega)}{2} \leq p + \omega + \frac{\delta(p + \omega)}{2}.$$

In other words, these imply that

$$I(p) \subseteq I(p + \omega).$$

In this case, $I(p)$ is *not* the largest closed interval anymore, a contradiction. Thus we have

$$p + \omega - \frac{\delta(p + \omega)}{2} > p - \frac{\delta(p)}{2}$$

which is equivalent to

$$\delta(p + \omega) - \delta(p) < 2\omega. \tag{7.49}$$

Similarly, if $p + \frac{\delta(p)}{2} \geq p + \omega + \frac{\delta(p+\omega)}{2}$, then we have

$$p + \omega - \frac{\delta(p + \omega)}{2} \geq p - \frac{\delta(p + \omega)}{2} \geq p + \omega - \frac{\delta(p)}{2} \geq p - \frac{\delta(p)}{2}.$$

In other words, these mean that

$$I(p + \omega) \subseteq I(p)$$

which is a contradiction again. Thus we have

$$p + \frac{\delta(p)}{2} < p + \omega + \frac{\delta(p + \omega)}{2}$$

which is equivalent to

$$\delta(p + \omega) - \delta(p) > -2\omega. \tag{7.50}$$

Combining the inequalities (7.49) and (7.50), we obtain

$$|\delta(p + \omega) - \delta(p)| < 2\omega$$

for every $\omega > 0$. By Definition 7.4 (Continuity), $\delta(p)$ is continuous. This proves our claim.

Since $\delta(p)$ is continuous on the compact set $[a, b]$, the Extreme Value Theorem ensures that there exists a $c \in [a, b]$ such that

$$0 < \delta(c) \leq \delta(p)$$

for every $p \in [a, b]$. Now this $\delta(c)$ evidently satisfies Definition 7.8 (Uniform Continuity).

We have completed the proof of the problem. ∎

Problem 7.25

⍟ ⍟ *Suppose that* $I = [0, 1]$ *and* $f : I \to I$ *is a continuous function such that* $f \circ f = f$. *Let* $E = \{x \in I \mid f(x) = x\}$. *Prove that* E *is connected.*

Proof. Let $y = f(x) \in f(I) \subseteq I$. Then the hypothesis shows that

$$f(f(x)) = f(x),$$

i.e., $f(y) = y$ or $y \in E$. In other words, we have $f(I) \subseteq E$. On the other hand, if $x \in E$, then we have $f(x) = x$ so that $x \in f(I)$. In conclusion, we have

$$E = f(I). \tag{7.51}$$

Since I is connected and f is continuous on I, Theorem 7.12 (Continuity and Connectedness) implies that $f(I)$ is also connected. Hence our desired result follows from the expression (7.51) immediately. This ends the proof of the problem. ∎

Problem 7.26

(⋆)(⋆) *Suppose that $f, g : [a, b] \to \mathbb{R}$ are continuous functions satisfying*

$$g(x) < f(x) \tag{7.52}$$

for all $x \in [a, b]$. Prove that there exists a $\delta > 0$ such that

$$g(x) + \delta < f(x)$$

for all $x \in [a, b]$.

Proof. Define the function $h : [a, b] \to \mathbb{R}$ by

$$h(x) = f(x) - g(x)$$

for all $x \in [a, b]$. Since f and g are continuous on $[a, b]$, h is also continuous on $[a, b]$. Thus it deduces from the Extreme Value Theorem that h attains a minimum value $h(q)$ *for some* $q \in [a, b]$, i.e., $h(x) \geq h(q)$ for all $[a, b]$. By the hypothesis (7.52), we know that $h(x) > 0$ for all $x \in [a, b]$ so that $h(q) > 0$. Let $\delta = \frac{h(q)}{2}$. Then we have

$$h(x) \geq h(q) > \delta$$

for all $x \in [a, b]$. This finished the proof of the problem. ∎

Problem 7.27

(⋆) *Let $f : X \to Y$ be a function from a metric space X to another metric space Y. If $f(p) = p$ for some $p \in X$, then we call p a **fixed point** of f.*
Suppose that $f : [a, b] \to \mathbb{R}$ is continuous, $f(a) \leq a$ and $f(b) \geq b$. Prove that f has a fixed point in $[a, b]$.

Proof. If $f(a) = a$ or $f(b) = b$, then f has a fixed point a or b and we are done. Thus without loss of generality, we may assume that $f(a) < a$ and $f(b) > b$. Define $g : [a, b] \to \mathbb{R}$ by

$$g(x) = f(x) - x$$

for $x \in [a,b]$. Since f is continuous on $[a,b]$, g is also continuous on $[a,b]$. Since we have $g(a) = f(a) - a < 0$ and $g(b) = f(b) - b > 0$, it follows from the Intermediate Value Theorem that there exists a $p \in (a,b)$ such that $g(p) = 0$, i.e.,

$$f(p) = p.$$

We finish the proof of the problem.   ■

Problem 7.28

$(\star)$ For each $n = 1, 2, \ldots$, suppose that $f_n : \mathbb{R} \to \mathbb{R}$ is defined by

$$f_n(x) = Ax^n + x^{n-1} + \cdots + x - 1,$$

where $A > 1$. Prove that $f_n(x)$ has a *unique* positive root α_n for each $n = 1, 2, \ldots$.

Proof. If $n = 1$, then $f_1(x) = Ax - 1$ which has a unique positive root $\frac{1}{A}$ obviously. Without loss of generality, we may assume that $n \geq 2$. By Remark 7.1, each $f_n(x)$ is continuous on $\mathbb{R}$. Furthermore, we have

$$f_n(0) = -1 < 0 \quad \text{and} \quad f_n(1) = A + (n-1) - 1 > n - 1 > 0.$$

By the Intermediate Value Theorem, we know that there exists a $\alpha_n \in (0,1)$ such that

$$f_n(\alpha_n) = 0.$$

If $x, y \in (0, +\infty)$ with $x > y$, then we have

$$
\begin{aligned}
f_n(x) - f_n(y) &= Ax^n + x^{n-1} + \cdots + x - 1 - (Ay^n + y^{n-1} + \cdots + y - 1) \\
&= A(x^n - y^n) + (x^{n-1} - y^{n-1}) + \cdots + (x - y) \\
&= (x - y)\Big[A(x^{n-1} + x^{n-2}y + \cdots + xy^{n-2} + y^{n-1}) \\
&\quad + (x^{n-2} + x^{n-3}y + \cdots + xy^{n-3} + y^{n-2}) + \cdots + (x + y) + 1 \Big] \quad (7.53)
\end{aligned}
$$

Since $x > y > 0$, we have $x - y > 0$ and every term in the expression (7.53) is positive. Thus $f_n(x) > f_n(y)$ and this means that f_n is strictly increasing on $(0, +\infty)$. Hence the positive root α_n of $f_n(x)$ is unique. This ends the proof of the problem.   ■

7.5 Discontinuity of Functions

Problem 7.29

$(\star)$ We say that a function f is **right continuous** at p if the **right-hand limit** of f at p exists and equals $f(p)$. (**Left continuous** at p can be defined similarly.) Construct a function which is *only* right continuous in its domain but unbounded.

Proof. Define $f : (-1, 0] \to \mathbb{R}$ by

$$f(x) = \begin{cases} \dfrac{1}{x}, & \text{if } x \in (-1, 0); \\[2mm] 0, & \text{if } x = 0. \end{cases}$$

We check Definition 7.13 (Left-hand Limits and Right-hand Limits). Let $p \in (-1, 0)$ and $\{x_n\} \subseteq (p, 0)$ be such that $x_n \to p$ as $n \to \infty$. Then we have

$$f(p) = \frac{1}{p} \quad \text{and} \quad f(x_n) = \frac{1}{x_n}$$

for each $n = 1, 2, \dots$ It is clear that $f(x_n) \to f(p)$, so f is right continuous at p.[g] Besides, f is not left continuous at 0 because the left-hand limit of f at 0 does not exit. In fact, f is unbounded on $(-1, 0]$ because

$$f(x_n) \to -\infty$$

for all sequences $\{x_n\} \subseteq (-1, 0)$ such that $x_n \to 0$ as $n \to \infty$, completing the proof of the problem. ∎

Problem 7.30

(⋆) Let $f : \mathbb{R} \to \mathbb{R}$ be the **greatest integer function**, i.e., $f(x) = [x]$. What are the type(s) of the discontinuities of f?

Proof. For each $n \in \mathbb{Z}$, if $n \le x < n + 1$, then we have

$$f(x) = [x] = n.$$

This implies that f is continuous on $(n, n + 1)$ for every $n \in \mathbb{Z}$. However, we have

$$f(n+) = \lim_{\substack{x \to n \\ x > n}} f(x) = n = f(n) \quad \text{and} \quad f(n-) = \lim_{\substack{x \to n \\ x < n}} f(x) = n - 1 \neq f(n).$$

By Definition 7.15 (Types of Discontinuity), f has simple discontinuities at every integer. This completes the proof of the problem. ∎

Problem 7.31

(⋆) Check the types of the discontinuities of the function $f : \mathbb{R} \to \mathbb{R}$ defined by

$$f(x) = \begin{cases} \sin \frac{1}{x}, & \text{if } x \neq 0; \\[2mm] 1, & \text{if } x = 0. \end{cases}$$

---

[g] In fact, f is continuous at every $p \in (-1, 0)$.

Proof. For $x \neq 0$, it is clear that f is continuous at x, so the only possible discontinuity of f is the origin. Let $x_n = \frac{1}{2n\pi}$ and $y_n = \frac{1}{(2n\pi + \frac{\pi}{2})}$. Then it is clear that $\{x_n\}, \{y_n\} \subseteq (0, +\infty)$, $x_n \to 0$ and $y_n \to 0$ as $n \to \infty$. Besides, we have

$$\lim_{n \to \infty} f(x_n) = \lim_{n \to \infty} \sin \frac{1}{x_n} = \lim_{n \to \infty} \sin 2n\pi = 0$$

and

$$\lim_{n \to \infty} f(y_n) = \lim_{n \to \infty} \sin \frac{1}{y_n} = \lim_{n \to \infty} \sin \left(2n\pi + \frac{\pi}{2}\right) = 1.$$

By Definition 7.13 (Left-hand Limits and Right-hand Limits), $f(0+)$ does not exist. By Definition 7.15 (Types of Discontinuity), we see that f has a discontinuity of the second kind at 0, completing the proof of the problem. ∎

7.6 Monotonic Functions

Problem 7.32

⋆ ⋆ *Suppose that $f : \mathbb{R} \to \mathbb{R}$ is a function. Prove that if f is strictly monotonic in $\mathbb{R}$, then f is one-to-one in $\mathbb{R}$.*

Proof. We just prove the case when the function f is strictly increasing. If $p, q \in \mathbb{R}$ and $p \neq q$, then either $p < q$ or $p > q$. If $p < q$, then since f is strictly increasing in $\mathbb{R}$, we have $f(p) < f(q)$. Similarly, if $p > q$, then since f is strictly increasing in $\mathbb{R}$, we have $f(p) > f(q)$. In both cases, we have

$$f(p) \neq f(q).$$

By definition (see §1.1.2), we conclude that f is one-to-one in $\mathbb{R}$, finishing the proof of the problem. ∎

Problem 7.33

⋆ ⋆ *Prove that if $f : \mathbb{R} \to \mathbb{R}$ is a continuous and one-to-one function, then it is strictly monotonic.*

Proof. Assume that f was not strictly monotonic. Then one can find $a < b < c$ such that either $f(a) < f(b) > f(c)$ or $f(a) > f(b) < f(c)$. Suppose that $f(a) < f(b) > f(c)$.[h] Then it implies that

$$f(b) > \max(f(a), f(c)).$$

Since f is continuous on $\mathbb{R}$, it is also continuous on $[a, b]$ and $[b, c]$. By the Intermediate Value Theorem, there exists $p \in (a, b)$ and $q \in (b, c)$ such that

$$f(p) = f(q).$$

However, $p < b < q$ which contradicts the hypothesis that f is one-to-one. Hence we have completed the proof of the problem. ∎

---

[h]The case for $f(a) > f(b) < f(c)$ is similar, so we omit the details here.

Remark 7.6

Combining Problems 7.32 and 7.33, for a continuous function $f : \mathbb{R} \to \mathbb{R}$, we establish the fact that f is one-to-one if and only if f is strictly monotonic.

Problem 7.34

$(\star)(\star)(\star)$ Let $f : (a, b) \to \mathbb{R}$. Suppose that for each $p \in (a, b)$, there exists a neighborhood $(p - \delta, p + \delta)$ of p in (a, b) such that f is decreasing in $(p - \delta, p + \delta)$. Show that f is decreasing in (a, b).

Proof. Assume that f was not decreasing in (a, b). Thus there exist $p, q \in (a, b)$ with $p < q$ such that $f(p) < f(q)$. By the hypothesis, we know that

$$[p, q] \subseteq \bigcup_{x \in [p,q]} (x - \delta_x, x + \delta_x) \subseteq (a, b),$$

where $\delta_x > 0$. Since $[p, q]$ is compact, we follow from Definition 4.12 (Compact Sets) that there are $x_1, x_2, \ldots, x_n \in [p, q]$ such that

$$[p, q] \subseteq \bigcup_{k=1}^{n} (x_k - \delta_{x_k}, x_k + \delta_{x_k}). \tag{7.54}$$

If $n = 1$, then we have $f(p) \geq f(q)$ which is a contradiction. Therefore, without loss of generality, we may assume that $n \geq 2$ and $p \leq x_1 < x_2 < \cdots < x_n \leq q$. In addition, we may assume further that there are no $i, j \in \{1, 2, \ldots, n\}$ such that

$$(x_i - \delta_{x_i}, x_i + \delta_{x_i}) \subseteq (x_j - \delta_{x_j}, x_j + \delta_{x_j}). \tag{7.55}$$

Otherwise, we may ignore the neighborhood $(x_i - \delta_{x_i}, x_i + \delta_{x_i})$ in the set relation (7.54). By the hypothesis, $p = x_1$ and $q = x_n$ so that

$$f(p) = f(x_1) \quad \text{and} \quad f(q) = f(x_n) \tag{7.56}$$

respectively. For each $k = 1, 2, \ldots, n - 1$, it is clear that

$$(x_k - \delta_{x_k}, x_k + \delta_{x_k}) \cap (x_{k+1} - \delta_{x_{k+1}}, x_{k+1} + \delta_{x_{k+1}}) \neq \varnothing.$$

Otherwise, there exists a $m \in \{1, 2, \ldots, n - 1\}$ such that

$$(x_m - \delta_{x_m}, x_m + \delta_{x_m}) \cap (x_{m+1} - \delta_{x_{m+1}}, x_{m+1} + \delta_{x_{m+1}}) = \varnothing.$$

By the relation (7.54) again, we can find a $p \in \{1, 2, \ldots, n - 1\} \setminus \{m, m + 1\}$ such that

$$(x_m + \delta_{x_m}, x_{m+1} + \delta_{x_{m+1}}) \subseteq (x_p - \delta_{x_p}, x_p + \delta_{x_p}). \tag{7.57}$$

If $x_p < x_m$, then we gain from the set relation (7.57) that $x_p + \delta_{x_p} > x_m + \delta_{x_m}$ and then it implies that

$$\delta_{x_p} > x_m - x_p + \delta_{x_m} > \delta_{x_m}.$$

In this case, we have

$$(x_m - \delta_{x_m}, x_m + \delta_{x_m}) \subseteq (x_p - \delta_{x_p}, x_p + \delta_{x_p})$$

which contradicts our assumption (7.55). If $x_{m+1} < x_p$, then the relation (7.57) evidently implies that $x_p - \delta_{x_p} < x_{m+1} - \delta_{x_{m+1}}$ but this shows that

$$\delta_{x_{m+1}} < \delta_{x_p} + x_{m+1} - x_p < \delta_{x_p}.$$

Therefore, we have

$$(x_{m+1} - \delta_{x_{m+1}}, x_{m+1} + \delta_{x_{m+1}}) \subseteq (x_p - \delta_{x_p}, x_p + \delta_{x_p}),$$

a contradiction again.

Now we let

$$y_k \in (x_k, x_k + \delta_{x_k}) \cap (x_{k+1} - \delta_{x_{k+1}}, x_{k+1})$$

for $k = 1, 2 \ldots, n-1$. Since f is decreasing in $(x_k - \delta_{x_k}, x_k + \delta_{x_k})$ and $(x_{k+1} - \delta_{x_{k+1}}, x_{k+1} + \delta_{x_{k+1}})$, we must have

$$f(x_k) \geq f(y_k) \geq f(x_{k+1}) \tag{7.58}$$

for every $k = 1, 2, \ldots, n-1$. Thus, by combining the equalities (7.56) and the inequalities (7.58), we conclude that

$$f(p) \geq f(x_1) \geq f(x_2) \geq \cdots \geq f(x_n) \geq f(q)$$

which is a contradiction. Hence f is decreasing in (a, b) and this completes the proof of the problem. ∎

Problem 7.35

⭐ ⭐ *Suppose that $f : [a, b] \to \mathbb{R}$ is monotonically decreasing and $a = p_0 < p_1 < \cdots < p_n = b$. Prove that*

$$\sum_{k=1}^{n-1} [f(p_k-) - f(p_k+)] \leq f(a+) - f(b-).$$

Proof. By Theorem 7.18 and Remark 7.3, we see that both $f(p_k+)$ and $f(p_k-)$ exist and

$$f(p_k-) \geq f(p_k) \geq f(p_k+) \tag{7.59}$$

for $k = 1, 2, \ldots, n$. Let $q_1, q_2, \ldots, q_n \in [a, b]$ be points such that

$$a = p_0 < q_1 < p_1 < q_2 < \cdots < q_k < p_k < q_{k+1} < \cdots < p_{n-1} < q_n < p_n = b.$$

Particularly, we have

$$q_k < p_k < q_{k+1}, \tag{7.60}$$

where $k = 1, 2, \ldots, n - 1$. Since f is monotonically decreasing on $[a, b]$, it follows from the inequalities (7.59) and (7.60) that

$$f(q_k) \geq f(p_k-) \geq f(p_k) \geq f(p_k+) \geq f(q_{k+1}) \tag{7.61}$$

which means that

$$f(q_k) - f(q_{k+1}) \geq f(p_k-) - f(p_k+)$$

for each $k = 1, 2, \ldots, n - 1$. Thus we deduce from the inequalities (7.61) and then using the decreasing property of f again to obtain that

$$
\begin{aligned}
\sum_{k=1}^{n-1}[f(p_k-) - f(p_k+)] &\leq \sum_{k=1}^{n-1}[f(q_k) - f(q_{k+1})] \\
&= [f(q_1) - f(q_2)] + [f(q_2) - f(q_3)] + \cdots + [f(q_{n-1}) - f(q_n)] \\
&= f(q_1) - f(q_n) \\
&\leq f(a+) - f(b-).
\end{aligned}
$$

We end the proof of the problem. ∎

CHAPTER **8**

Differentiation

In this section, we briefly review the definitions, notations and properties of derivatives of real functions defined on intervals (a, b) or $[a, b]$. The main references for this part are [2, Chap. 4], [3, Chap. 5], [5, Chap. 6] and [13, Chap. 5].

8.1 Fundamental Concepts

8.1.1 Definitions and Notations

Definition 8.1. *Suppose that $f : [a, b] \to \mathbb{R}$ and $x \in [a, b]$. Define the function $\phi : [a, b] \backslash \{x\} \to \mathbb{R}$ by*

$$\phi(t) = \frac{f(t) - f(x)}{t - x}$$

and define the limit (if it exists)

$$f'(x) = \lim_{t \to x} \phi(t) = \lim_{t \to x} \frac{f(t) - f(x)}{t - x}. \tag{8.1}$$

Equivalently, the number $f'(x)$ given in (8.1) can also be defined as

$$f'(x) = \lim_{h \to 0} \frac{f(x + h) - f(x)}{h}. \tag{8.2}$$

The new function f' is called the **derivative** of f whose domain is the subset $E \subseteq (a, b)$ at which the limit (8.1) exists and the process of obtaining f' from f is called **differentiation**. If f' is defined at x, then it is said that f is **differentiable** at x. Similarly, we say f is **differentiable** in E. At the end-points a and b, we consider the **right-hand derivative** $f'(a+)$ and the **left-hand derivative** $f'(b-)$ respectively.[a]

---

[a]They are

$$f'(a+) = \lim_{\substack{t \to a \\ t > a}} \frac{f(t) - f(a)}{t - a} \quad \text{and} \quad f'(b-) = \lim_{\substack{t \to b \\ t < b}} \frac{f(t) - f(b)}{t - b}.$$

> **Remark 8.1**
>
> We assume the reader is familiar with the derivatives of some well-known functions such as constant functions, linear functions, power functions, polynomials, rational functions, trigonometric functions, logarithmic/exponential functions and etc.

8.1.2 Elementary Properties of Derivatives

Theorem 8.2. *Suppose that* $f : [a, b] \to \mathbb{R}$. *If* f *is differentiable at* $p \in [a, b]$, *then* f *is continuous at* p.

The next result describes the usual operations for derivatives of the sum, difference, product and quotient of two differentiable functions.

Theorem 8.3 (Operations of Differentiable Functions). *Suppose that* $f, g : [a, b] \to \mathbb{R}$ *and they are differentiable at* $p \in [a, b]$. *Then we have the following formulas:*

(a) $(f \pm g)'(p) = f'(p) \pm g'(p)$;

(b) $(f \cdot g)'(p) = f(p)g'(p) + f'(p)g(p)$;

(c) $\left(\dfrac{f}{g} \right)'(p) = \dfrac{g(p)f'(p) - g'(p)f(p)}{g^2(p)}$.

For composition of differentiable functions, we have the famous **Chain Rule**:

Theorem 8.4 (Chain Rule). *Suppose that* $f : [a, b] \to \mathbb{R}$ *is continuous and* $f'(p)$ *exists at some point* $p \in [a, b]$. *Furthermore, if* $g : I \to \mathbb{R}$, $f([a, b]) \subseteq I$ *and* g *is differentiable at* $f(p)$. *If* $h : [a, b] \to \mathbb{R}$ *is defined by*

$$h(t) = g(f(t)),$$

then h *is differentiable at* p *and its derivative is given by*

$$h'(p) = g'(f(p)) \times f'(p).$$

8.1.3 Local Maxima/Minima and Zero Derivatives

Definition 8.5. *Let* X *be a metric space with metric* d *and* $f : X \to \mathbb{R}$ *be a function. Then* f *is said to have a **local maximum** at* $p \in X$ *if there is a* $\delta > 0$ *such that*

$$f(p) \geq f(x)$$

for all $x \in X$ *with* $d(x, p) < \delta$.

Local minima can be defined similarly. A point p is called a **local extreme** of f if it is either a local maximum or a local minimum of f.

> **Remark 8.2**
>
> It is clear that if f has an absolute maximum/minimum at p, then p is also a local maximum/minimum. However, the converse is false.

Theorem 8.6 (Fermat's Theorem). *Suppose that $f : [a,b] \to \mathbb{R}$ has a local maximum/minimum at a point $p \in (a,b)$ and $f'(p)$ exists. Then we have*

$$f'(p) = 0.$$

8.1.4 The Mean Value Theorem and the Intermediate Value Theorem for Derivatives

Theorem 8.7 (Rolle's Theorem). *Suppose that $f : [a,b] \to \mathbb{R}$ is continuous on $[a,b]$ and differentiable in (a,b). If $f(a) = f(b)$, then there exists a point $p \in (a,b)$ such that*

$$f'(p) = 0.$$

Theorem 8.8 (The Generalized Mean Value Theorem). *Suppose that $f, g : [a,b] \to \mathbb{R}$ are continuous on $[a,b]$ and differentiable in (a,b). Then there exists a point $p \in (a,b)$ such that*

$$[f(b) - f(a)]g'(p) = [g(b) - g(a)]f'(p).$$

We note that the above theorem is also called the **Cauchy Mean Value Theorem**.

The Mean Value Theorem for Derivatives. *If $f : [a,b] \to \mathbb{R}$ is continuous on $[a,b]$ and differentiable in (a,b), then there exists a point $p \in (a,b)$ such that*

$$f(b) - f(a) = (b - a)f'(p).$$

Geometrically, the above theorem states that a sufficiently "smooth" curve joining two points A and B in the plane has *at least* one tangent line with the **same** slope as the chord AB. The following result is an immediate consequence of the Mean Value Theorem for Derivatives.

Theorem 8.9. *Suppose that $f : [a,b] \to \mathbb{R}$ is continuous on $[a,b]$ and differentiable in (a,b).*

(a) *If $f'(x) \geq 0$ for all $x \in (a,b)$, then f is **monotonically increasing** on $[a,b]$.*

(b) *If $f'(x) \leq 0$ for all $x \in (a,b)$, then f is **monotonically decreasing** on $[a,b]$.*

(c) *If $f'(x) = 0$ for all $x \in (a,b)$, then $f \equiv c$ on $[a,b]$ for some constant c on $[a,b]$.*

> **Remark 8.3**
>
> If the inequalities in Theorem 8.9(a) and (b) are strict, then f is called **strictly increasing** on $[a,b]$ or **strictly decreasing** on $[a,b]$ respectively.

Similar to the Intermediate Value Theorem, we have an analogous result for the derivative of a function f.[b]

---

[b]This is also called **Darboux's Theorem**.

The Intermediate Value Theorem for Derivatives. *Suppose that* $f : [a, b] \to \mathbb{R}$ *is differentiable in* $[a, b]$. *If* λ *is a number between* $f'(a)$ *and* $f'(b)$, *then there exists a* $p \in (a, b)$ *such that*

$$f'(p) = \lambda.$$

▎ 8.1.5 L'Hôspital's Rule

Sometimes it is difficult to evaluate limits in the form

$$\lim_{x \to p} \frac{f(x)}{g(x)}$$

because $f(p) \to 0$ and $g(p) \to 0$ as $x \to p$. In this case, the limit of the quotient $\frac{f(x)}{g(x)}$ is said to be an **indeterminate form**. Symbolically, we use $\frac{0}{0}$ to denote this situation. Other indeterminate forms consist of the cases

$$\frac{\infty}{\infty}, \quad 0 \cdot \infty, \quad 0^0, \quad 1^\infty, \quad \infty^0 \quad \text{and} \quad \infty - \infty.$$

To compute limits in one of the above indeterminate forms, we may need the help of the so-called **L'Hôspital's Rule** and it is stated as follows:

Theorem 8.10 (L'Hôspital's Rule). *Suppose that* $-\infty \leq a < b < +\infty$, $f, g : (a, b) \to \mathbb{R}$ *are differentiable in* (a, b), $g'(x) \neq 0$ *on* (a, b) *and*

$$\lim_{\substack{x \to a \\ x > a}} f(x) = \lim_{\substack{x \to a \\ x > a}} g(x) = 0.$$

Then we must have

$$\lim_{\substack{x \to a \\ x > a}} \frac{f(x)}{g(x)} = \lim_{\substack{x \to a \\ x > a}} \frac{f'(x)}{g'(x)} = L,$$

where $L \in [-\infty, +\infty]$.

> **Remark 8.4**
>
> In the preceding result, only the right-hand limit of $\frac{f(x)}{g(x)}$ at a is presented. In fact, similar results for the left-hand limit of $\frac{f(x)}{g(x)}$ at a or two-sided limits of a can be stated and treated.

The following result is very similar to Theorem 8.10 (L'Hôspital's Rule), but the main difference is that it considers the case where the denominator tends to ∞ from the right-hand side of a.

Theorem 8.11. *Suppose that* $-\infty \leq a < b < \infty$, $f, g : (a, b) \to \mathbb{R}$ *are differentiable on* (a, b), $g'(x) \neq 0$ *on* (a, b) *and*

$$\lim_{\substack{x \to a \\ x > a}} g(x) = \pm\infty.$$

Then we must have

$$\lim_{\substack{x \to a \\ x > a}} \frac{f(x)}{g(x)} = \lim_{\substack{x \to a \\ x > a}} \frac{f'(x)}{g'(x)} = L,$$

where $L \in [-\infty, +\infty]$.

∎ 8.1.6 Higher Order Derivatives and Taylor's Theorem

Definition 8.12. *Suppose that f is a differentiable function in (a, b). If f', in turn, is defined on an open interval and if f' is itself differentiable, then we denote the first derivative of f' by f'', called the **second derivative** of f. Similarly, the **nth derivative** of f,[c] denoted by $f^{(n)}$, can be defined to be the first derivative of the function $f^{(n-1)}$ in an open interval. The function f itself can be written as $f^{(0)}$.*

We notice that if $f^{(n)}$ exists at a point $x \in (a, b)$, then the function $f^{(n-1)}(t)$ must exist for all $t \in (x - \delta, x + \delta) \subseteq (a, b)$ for some $\delta > 0$. Therefore, $f^{(n-1)}$ is also differentiable at x. Since $f^{(n-1)}$ exists on the open interval $(x - \delta, x + \delta)$, the function $f^{(n-2)}$ must be differentiable in $(x - \delta, x + \delta)$.

The following result is known as **Leibniz's rule** which generalizes the product rule of two differentiable functions.

Leibniz's Rule. *Suppose that f and g have nth derivatives at x. Then we have*

$$(fg)^{(n)}(x) = \sum_{k=0}^{n} C_k^n f^{(n-k)}(x) g^{(k)}(x),$$

where C_k^n is the binomial coefficient.

Furthermore, with the help of higher order derivatives, we can talk about the approximation of a function f by a particular polynomial at a point p. This is the main ingredient of the famous Taylor's Theorem.

Taylor's Theorem. *Let $n \in \mathbb{N}$. Suppose that $f : [a, b] \to \mathbb{R}$ satisfies the conditions:*

- *$f^{(n-1)}$ is continuous on $[a, b]$ and*

- *$f^{(n)}$ exists in (a, b).*

Then for every distinct points $x, p \in [a, b]$, there exists a point ξ between x and p such that[d]

$$f(x) = \underbrace{\sum_{k=0}^{n-1} \frac{f^{(k)}(p)}{k!} (x - p)^k}_{Taylor's\ polynomial} + \underbrace{\frac{f^{(n)}(\xi)}{n!} (x - p)^n}_{Error\ term}.$$

Remark 8.5

We notice that the domain $[a, b]$ in Taylor's Theorem can be replaced by any open interval containing the point p.

---

[c]Or the derivative of order n of f.
[d]That is either $\xi \in (p, x)$ or $\xi \in (x, p)$.

▍ 8.1.7 Convexity and Derivatives

The concepts of convexity and convex functions play an important role in different areas of mathematics. Particularly, they are useful and important in the study of optimization problems.

Definition 8.13 (Convex Functions). *A function $f : (a, b) \to \mathbb{R}$ is called **convex** on (a, b) if for every t satisfying $0 \le t \le 1$, we have*

$$f(tx_1 + (1-t)x_2) \le tf(x_1) + (1-t)f(x_2) \tag{8.3}$$

whenever $x_1, x_2 \in (a, b)$.[e]

Geometrically, if f is convex on (a, b), then, for every $x_1, x_2 \in (a, b)$, the chord joining the two points $(x_1, f(x_1))$ and $(x_2, f(x_2))$ lies above the **graph** of f. One of the important properties of convex functions is the following result:

Theorem 8.14. *If $f : [a, b] \to \mathbb{R}$ is convex, then f is continuous on (a, b).*

However, a convex function f needs not be differentiable at a point, but if f has the first derivative or the second derivative, then we can establish some connections between f and its derivatives.

Theorem 8.15. *Suppose that f is a real differentiable function defined in (a, b). Then f is convex on (a, b) if and only if f' is **monotonically increasing** on (a, b).*

Theorem 8.16. *Suppose that f is a twice differentiable real function defined in (a, b). Then f is convex on (a, b) if and only if $f''(x) \ge 0$ for all $x \in (a, b)$.*

Remark 8.6

A function $f : (a, b) \to \mathbb{R}$ is called **concave** on (a, b) if the sign "$\le$" in the inequality (8.3) is replaced by "$\ge$". In this case, the chord joining the two points $(x_1, f(x_1))$ and $(x_2, f(x_2))$ lies below the graph of f.

▍ 8.2 Properties of Derivatives

Problem 8.1

$(\star)$ *Suppose that $f : \mathbb{R} \to \mathbb{R}$ is continuous at p. Define $g : \mathbb{R} \to \mathbb{R}$ by*

$$g(x) = (x - p)f(x)$$

for all $x \in \mathbb{R}$. Prove that $g'(p)$ exists.

Proof. By the definition (8.2), we have

$$g'(p) = \lim_{h \to 0} \frac{g(p+h) - g(p)}{h} = \lim_{h \to 0} \frac{hf(p+h) - 0}{h} = \lim_{h \to 0} f(p+h).$$

---

[e]If the inequality is strict, then f is said to be **strictly convex** on (a, b).

Since f is continuous at p, it follows from Theorem 7.5 that

$$\lim_{h \to 0} f(p + h) = f(p)$$

so that

$$g'(p) = f(p).$$

This ends the proof of the problem. ∎

Problem 8.2

⋆ Let $a > 1$ and $f : \mathbb{R} \to \mathbb{R}$ be a function defined by

$$f(x) = \begin{cases} x^a \sin \frac{1}{x}, & \text{if } x \neq 0; \\ 0, & \text{if } x = 0. \end{cases}$$

Prove that f is differentiable at 0.

Proof. By Definition 8.1, we have

$$\lim_{t \to 0} \frac{f(t) - f(0)}{t - 0} = \lim_{t \to 0} t^{a-1} \sin \frac{1}{t}.$$

Since $a > 1$ and $|\sin \frac{1}{x}| \leq 1$ for every $x \in \mathbb{R}$, we know that

$$0 \leq \left| t^{a-1} \sin \frac{1}{t} \right| \leq |t|^{a-1} \tag{8.4}$$

Apply Theorem 7.3 (Squeeze Theorem for Limits of Functions) to the inequalities (8.4) to get

$$\lim_{t \to 0} t^{a-1} \sin \frac{1}{t} = 0.$$

Thus f is differentiable at 0 and $f'(0) = 0$. This finishes the proof of the problem. ∎

Problem 8.3

⋆ Let $m, n \in \mathbb{Z}$ and $f : \mathbb{R} \to \mathbb{R}$ be a function defined by

$$f(x) = \begin{cases} x^n \sin \frac{1}{x^m}, & \text{if } x \neq 0; \\ 0, & \text{if } x = 0. \end{cases}$$

Suppose that $n \geq m + 1$ and $n \geq 1$. Prove that there exists a $\delta > 0$ such that $f'(x)$ is bounded on the segment $(-\delta, \delta)$.

Proof. By applying Theorem 8.3 (Operations of Differentiable Functions) repeatedly, we have

$$f'(x) = nx^{n-1} \sin \frac{1}{x^m} - mx^{n-m-1} \cos \frac{1}{x^m} \tag{8.5}$$

if $x \in (-\delta, \delta)$ and $x \neq 0$. Since $n \geq 1$ and $n \geq m + 1$, we have x^{n-1} and x^{n-m-1} are bounded on $(-1, 1)$. It is well-known that $\sin x$ and $\cos x$ are bounded on $\mathbb{R}$. Combining these facts and the derivative (8.5), we know that $f'(x)$ is bounded on $(-1, 1) \setminus \{0\}$.

To check the remaining case $f'(0)$, we notice from the definition (8.2) and an application of Theorem 7.3 (Squeeze Theorem for Limits of Functions) that

$$f'(0) = \lim_{h \to 0} \frac{f(0+h) - f(0)}{h} = \lim_{h \to 0} h^{n-1} \sin \frac{1}{h^m} = 0.$$

Hence we conclude that $f'(x)$ is bounded on $(-1, 1)$, completing the proof of the problem. ∎

Problem 8.4

($\star$) *Prove that the function $f : \mathbb{R} \to \mathbb{R}$ defined by*

$$f(x) = \begin{cases} x^3, & \text{if } x \in \mathbb{Q}; \\ \\ 0, & \text{if } x \in \mathbb{R} \setminus \mathbb{Q} \end{cases}$$

is differentiable only at 0.

Proof. We note that

$$\frac{f(0+h) - f(0)}{h} = \frac{f(h)}{h} = \begin{cases} h^2, & \text{if } h \in \mathbb{Q}; \\ \\ 0, & \text{if } h \in \mathbb{R} \setminus \mathbb{Q}. \end{cases}$$

By the definition (8.2), we have

$$f'(0) = \lim_{h \to 0} \frac{f(0+h) - f(0)}{h}$$

$$= \begin{cases} \lim_{h \to 0} \dfrac{h^2}{h}, & \text{if } h \in \mathbb{Q}; \\ \\ \lim_{h \to 0} \dfrac{0}{h}, & \text{if } h \in \mathbb{R} \setminus \mathbb{Q} \end{cases}$$

$$= 0.$$

In other words, f is differentiable at 0.

Suppose that $x \in \mathbb{Q} \setminus \{0\}$. Then we deduce from Remark 2.1 that we can find a sequence $\{x_n\} \subseteq \mathbb{R} \setminus (\mathbb{Q} \cup \{x\})$ such that $x_n \to x$ as $n \to \infty$. By this sequence, we have

$$\lim_{n \to \infty} \frac{f(x_n) - f(x)}{x_n - x} = \lim_{n \to \infty} \frac{-x^3}{x_n - x}. \tag{8.6}$$

Since $x \neq 0$ and $x_n \to x$ as $n \to \infty$, the limit (8.6) is either $-\infty$ or $+\infty$ and this implies that f is not differentiable at every $x \in \mathbb{Q} \setminus \{0\}$. Similarly, if $x \in \mathbb{R} \setminus \mathbb{Q}$ and $x \neq 0$, then it follows from Theorem 2.2 (Density of Rationals) that there exists a sequence $\{y_n\} \subseteq \mathbb{Q} \setminus \{x, 0\}$ such that $y_n \to x$ as $n \to \infty$. By this sequence, we have

$$\lim_{n \to \infty} \frac{f(y_n) - f(x)}{y_n - x} = \lim_{n \to \infty} \frac{y_n^3}{y_n - x}. \tag{8.7}$$

Since $y_n \neq 0$ for all $n \in \mathbb{N}$, the limit (8.7) is either $-\infty$ or $+\infty$. Hence we conclude that f is not differentiable at every $x \in \mathbb{R} \setminus \{0\}$. We have completed the proof of the problem. ∎

Problem 8.5

⊛ Prove that the function $f : \mathbb{R} \to \mathbb{R}$ defined by

$$f(x) = x|x|$$

is differentiable in $\mathbb{R}$.

Proof. It is clear that

$$f(x) = \begin{cases} x^2, & \text{if } x \geq 0; \\ -x^2, & \text{if } x < 0. \end{cases}$$

Therefore, f is differentiable at every $x \in \mathbb{R} \setminus \{0\}$. We have to check the differentiability of f at the point 0. By the definition (8.2), we have

$$f'(0+) = \lim_{\substack{h \to 0 \\ h > 0}} \frac{f(0+h) - f(0)}{h} = \lim_{\substack{h \to 0 \\ h > 0}} h = 0$$

and

$$f'(0-) = \lim_{\substack{h \to 0 \\ h < 0}} \frac{f(0+h) - f(0)}{h} = \lim_{\substack{h \to 0 \\ h < 0}} -h = 0.$$

By Definition 8.1, f is differentiable at 0. Hence f is differentiable in $\mathbb{R}$, completing the proof of the problem. ∎

Problem 8.6

⊛ ⊛ Suppose that $f : \mathbb{R} \to \mathbb{R}$ is differentiable at p. Let $\{x_n\}$ and $\{y_n\}$ be increasing and decreasing sequences in $\mathbb{R} \setminus \{p\}$ respectively, both of them converge to p. Prove that

$$f'(p) = \lim_{n \to \infty} \frac{f(x_n) - f(y_n)}{x_n - y_n}.$$

Proof. Let

$$g(x) = \frac{f(x) - f(p)}{x - p} - f'(p). \tag{8.8}$$

Since $f'(p)$ exists, it follows from Definition 8.1 and Theorem 8.3 (Operations of Differentiable Functions) that

$$\lim_{x \to p} g(x) = f'(p) - f'(p) = 0. \tag{8.9}$$

Rewrite the expression (8.8) as

$$f(x) = f(p) + [f'(p) + g(x)](x - p)$$

which implies that

$$\frac{f(x_n) - f(y_n)}{x_n - y_n} = \frac{[f'(p) + g(x_n)](x_n - p) - [f'(p) + g(y_n)](y_n - p)}{x_n - y_n}$$

$$= \frac{f'(p)(x_n - y_n) + g(x_n)(x_n - p) - g(y_n)(y_n - p)}{x_n - y_n}$$

$$= f'(p) + g(x_n)\left(\frac{x_n - p}{x_n - y_n}\right) - g(y_n)\left(\frac{y_n - p}{x_n - y_n}\right)$$

for all $n \in \mathbb{N}$. Thus we have

$$\left|\frac{f(x_n) - f(y_n)}{x_n - y_n} - f'(p)\right| = \left|g(x_n)\left(\frac{x_n - p}{x_n - y_n}\right) - g(y_n)\left(\frac{y_n - p}{x_n - y_n}\right)\right|$$

$$\leq |g(x_n)| \times \left|\frac{x_n - p}{x_n - y_n}\right| + |g(y_n)| \times \left|\frac{y_n - p}{x_n - y_n}\right| \qquad (8.10)$$

for all $n \in \mathbb{N}$. Since $\{x_n\}$ is increasing in $\mathbb{R} \setminus \{p\}$ and converging to p, we have $x_n < p$ for all $n \in \mathbb{N}$. Similarly, we have $p < y_n$ for all $n \in \mathbb{N}$. Therefore, we have $x_n < p < y_n$ for all $n \in \mathbb{N}$ and this implies that

$$\left|\frac{x_n - p}{x_n - y_n}\right| \leq 1 \quad \text{and} \quad \left|\frac{y_n - p}{x_n - y_n}\right| \leq 1. \qquad (8.11)$$

By the inequalities (8.10) and (8.11), we obtain

$$\left|\frac{f(x_n) - f(y_n)}{x_n - y_n} - f'(p)\right| \leq |g(x_n)| + |g(y_n)| \qquad (8.12)$$

for all $n \in \mathbb{N}$. By the limit (8.9), Theorems 5.6 (Squeeze Theorem for Convergent Sequences) and 7.1, we see that

$$\lim_{n \to \infty} |g(x_n)| = \lim_{n \to \infty} |g(y_n)| = 0.$$

Hence we deduce from the inequality (8.12) that

$$\lim_{n \to \infty} \left|\frac{f(x_n) - f(y_n)}{x_n - y_n} - f'(p)\right| = 0.$$

By definition (see §2.1.2), we have $-|a| \leq a \leq |a|$ and so

$$\lim_{n \to \infty} \frac{f(x_n) - f(y_n)}{x_n - y_n} = f'(p)$$

as required. This completes the proof of the problem. ∎

Problem 8.7

(⋆) Suppose that $f : \mathbb{R} \to \mathbb{R}$ is differentiable at p. Prove that

$$\lim_{x \to p} \frac{p^n f(x) - x^n f(p)}{x - p} = p^n f'(p) - n f(p) p^{n-1}.$$

Proof. We notice that

$$\frac{p^n f(x) - x^n f(p)}{x - p} = \frac{p^n f(x) - p^n f(p) + p^n f(p) - x^n f(p)}{x - p}$$

$$= p^n \cdot \frac{f(x) - f(p)}{x - p} - f(p) \cdot \frac{x^n - p^n}{x - p}. \tag{8.13}$$

Since

$$\lim_{x \to p} \frac{f(x) - f(p)}{x - p} = f'(p) \quad \text{and} \quad \lim_{x \to p} \frac{x^n - p^n}{x - p} = np^{n-1},$$

we deduce from the expression (8.13) that

$$\lim_{x \to p} \frac{p^n f(x) - x^n f(p)}{x - p} = p^n \lim_{x \to p} \frac{f(x) - f(p)}{x - p} - f(p) \lim_{x \to p} \frac{x^n - p^n}{x - p}$$

$$= p^n f'(p) - n f(p) p^{n-1}$$

which is our desired result. This finishes the proof of the problem. ∎

Problem 8.8

(⋆) *Is the converse of Theorem 8.6 (Fermat's Theorem) true? Furthermore, construct a function f having a local minimum at 0 but $f'(0) \neq 0$.*

Proof. For example, we consider the function $f(x) = x^3$ whose derivative is $f'(x) = 3x^2$ so that $f'(0) = 0$ if and only if $x = 0$. However, for every $\epsilon > 0$, we can always find $p, q \in (-\epsilon, \epsilon)$ such that

$$p^3 < 0 < q^3$$

which means that 0 is *not* a local maximum or a local minimum of f.

Furthermore, note that the function $f : \mathbb{R} \to \mathbb{R}$ defined by $f(x) = |x|$ has the absolute (and hence local) minimum at $x = 0$, but it is *not* differentiable at 0 because

$$f'(0+) = \lim_{\substack{t \to 0 \\ t > 0}} \frac{f(t) - f(0)}{t - 0} = 1 \quad \text{and} \quad f'(0-) = \lim_{\substack{t \to 0 \\ t < 0}} \frac{f(t) - f(0)}{t - 0} = -1.$$

Thus we have completed the proof of the problem. ∎

Problem 8.9

(⋆)(⋆) *Suppose that $f : [0, 1] \to \mathbb{R}$ is differentiable in $[0, 1]$ and there is no $p \in [0, 1]$ such that $f(p) = f'(p) = 0$. Prove that the set*

$$Z = \{x \in [0, 1] \mid f(x) = 0\}$$

is finite.

Proof. Assume that there was an infinite sequence $\{p_n\} \subseteq [0,1]$ such that $f(p_n) = 0$. Since $\{p_n\}$ is bounded, Problem 5.25 (The Bolzano-Weierstrass Theorem) shows that $\{p_n\}$ has a convergent subsequence $\{p_{n_k}\}$. Without loss of generality, we may assume that $\{p_n\}$ converges to p. Since $[0,1]$ is compact (or closed), we have $p \in [0,1]$. Since f is differentiable in $[0,1]$, it is continuous on $[0,1]$ by Theorem 8.2. Therefore, it follows from Theorem 7.5 that

$$f(p) = f\left(\lim_{n\to\infty} p_n\right) = \lim_{n\to\infty} f(p_n) = 0. \tag{8.14}$$

Recall that $f(p_n) = 0$ for all $n = 1, 2, \ldots$, we derive by using Theorem 7.5 and the value (8.14) that

$$f'(p) = \lim_{x\to p} \frac{f(x) - f(p)}{x - p} = \lim_{n\to\infty} \frac{f(p_n) - f(p)}{p_n - p} = 0$$

which contradicts the hypothesis. This ends the proof of the problem. ∎

Problem 8.10

⊛ *Suppose that $f : \mathbb{R} \to \mathbb{R}$ is differentiable in $\mathbb{R}$. If the following formula*

$$f(x + h) = f(x) + hf'(x) \tag{8.15}$$

holds for every $x, h \in \mathbb{R}$, prove that there exist constants A and B such that

$$f(x) = Ax + B$$

for every $x \in \mathbb{R}$.

Proof. Take $h = y - 1$ and $x = 1$ in the formula (8.15), where $y \in \mathbb{R}$. Then it deduces that

$$f(1 + (y - 1)) = f(1) + (y - 1)f'(1)$$
$$f(y) = f'(1)y + [f(1) - f'(1)].$$

Thus we have

$$A = f'(1) \quad \text{and} \quad B = f(1) - f'(1).$$

Hence we have completed the proof of the problem. ∎

8.3   The Mean Value Theorem for Derivatives

Problem 8.11

⊛ ⊛ *Suppose that $f : [0,1] \to \mathbb{R}$ is continuous and $f(0) = 0$. Furthermore, $f(x)$ is differentiable for every $x \in (0,1)$ and*

$$0 \le f'(x) \le 2f(x) \tag{8.16}$$

on $(0,1)$. Prove that $f \equiv 0$ on $[0,1]$.

Proof. Define $F : [0, 1] \to \mathbb{R}$ by

$$F(x) = e^{-2x} f(x)$$

for every $x \in [0, 1]$. By Theorem 8.3 (Operations of Differentiable Functions), we see that

$$F'(x) = e^{-2x}[f'(x) - 2f(x)]$$

for every $x \in (0, 1)$. By the hypothesis (8.16), we have

$$F'(x) \le 0$$

for all $x \in (0, 1)$. By Theorem 8.9(b), we derive that F is monotonically decreasing on $[0, 1]$. Since $F(0) = e^0 f(0) = 0$, we conclude that

$$F(x) \le 0 \tag{8.17}$$

on $[0, 1]$. By the hypothesis (8.16) again, we have $f(x) \ge 0$ for every $(0, 1)$. Thus it follows from this and the continuity of f that

$$f(x) \ge 0$$

on $[0, 1]$. Therefore, we have

$$F(x) \ge 0 \tag{8.18}$$

on $[0, 1]$. Combining the inequalities (8.17) and (8.18), we obtain $F(x) \equiv 0$ on $[0, 1]$ which is equivalent to

$$f(x) \equiv 0$$

on $[0, 1]$. This completes the proof of the problem.   ∎

Problem 8.12

⋆ ⋆ *Suppose that $f : [a, b] \to \mathbb{R}$ is continuous on $[a, b]$ and differentiable in (a, b). Furthermore, the derivative $f'(x)$ is bounded on (a, b). Prove that f is uniformly continuous on (a, b).*

Proof. Since $f'(x)$ is bounded on (a, b), there exists a $M > 0$ such that

$$|f'(x)| \le M \tag{8.19}$$

for all $x \in (a, b)$. By the Mean Value Theorem for Derivatives, we have

$$f(x) - f(y) = f'(p)(x - y) \tag{8.20}$$

for some $p \in (x, y) \subseteq (a, b)$. For each $\epsilon > 0$, we take $\delta = \frac{\epsilon}{M}$. Therefore, if $x, y \in (a, b)$ and $|x - y| < \delta$, then we deduce from the inequality (8.19) and the expression (8.20) that

$$|f(x) - f(y)| = |f'(p)||x - y| \le M|x - y| < M\delta < \epsilon.$$

Hence f is uniformly continuous on (a, b) by Definition 7.8 (Uniform Continuity). This finishes the proof of the problem.   ∎

Problem 8.13

⋆ ⋆ *Suppose that* $f : (a,b) \to \mathbb{R}$ *is differentiable and its derivative* f' *is uniformly continuous on* (a,b). *Prove that*

$$\lim_{n \to \infty} n\left[f(x) - f\left(x - \frac{1}{n}\right)\right] = f'(x)$$

for every $x \in (a,b)$.

Proof. Since f' is uniformly continuous on (a,b), for every $\epsilon > 0$, there exists a $\delta > 0$ such that

$$|f'(x) - f'(y)| < \epsilon \tag{8.21}$$

for every $x, y \in (a,b)$ with $|x - y| < \delta$. By Theorem 2.1 (The Archimedean Property), there exists a positive integer N such that $N\delta > 1$. Then for $x \in (a,b)$, we choose $n \geq N$ such that $x - \frac{1}{n} > a$. Therefore, we have

$$\left(x - \frac{1}{n}, x\right) \subseteq (a,b) \quad \text{and} \quad \left|x - \frac{1}{n} - x\right| = \frac{1}{n} < \delta. \tag{8.22}$$

By this, we obtain from the inequality (8.21) that

$$|f'(p) - f'(x)| < \epsilon \tag{8.23}$$

for all $p \in (x - \frac{1}{n}, x) \subseteq (a,b)$.

Since f is differentiable in (a,b), it is continuous on $[x - \frac{1}{n}, x] \subset (a,b)$. Applying the Mean Value Theorem for Derivatives, we conclude that

$$\left|n\left[f(x) - f\left(x - \frac{1}{n}\right)\right] - f'(x)\right| = \left|\frac{f(x) - f(x - \frac{1}{n})}{x - (x - \frac{1}{n})} - f'(x)\right| = |f'(p_n) - f'(x)| \tag{8.24}$$

for some $p_n \in (x - \frac{1}{n}, x)$ for those n satisfying the requirements (8.22). By the inequalities (8.23) and (8.24), we have

$$\left|n\left[f(x) - f\left(x - \frac{1}{n}\right)\right] - f'(x)\right| < \epsilon$$

for some $p_n \in (x - \frac{1}{n}, x)$ for those n satisfying the requirements (8.22). By Definition 8.1, we get the desired result that

$$\lim_{n \to \infty} n\left[f(x) - f\left(x - \frac{1}{n}\right)\right] = f'(x)$$

for every $x \in (a,b)$. This ends the proof of the problem. ∎

Problem 8.14

⋆ ⋆ *Prove that for* $x, y > 0$ *and* $0 < \alpha < \beta$, *we have*

$$(x^\alpha + y^\alpha)^{\frac{1}{\alpha}} > (x^\beta + y^\beta)^{\frac{1}{\beta}}. \tag{8.25}$$

Proof. If $x = y$, then we have

$$(x^\alpha + y^\alpha)^{\frac{1}{\alpha}} = 2^{\frac{1}{\alpha}} x \quad \text{and} \quad (x^\beta + y^\beta)^{\frac{1}{\beta}} = 2^{\frac{1}{\beta}} x.$$

Since $2^{\frac{1}{\alpha}} > 2^{\frac{1}{\beta}}$, the inequality holds in this case.

Without loss of generality, we may assume that $x \neq y$. Define the function $f : (0, \infty) \to \mathbb{R}$ by

$$f(t) = (x^t + y^t)^{\frac{1}{t}}.$$

By Theorem 8.4 (Chain Rule), we have

$$f'(t) = \frac{f(t)}{t^2(x^t + y^t)} \left[x^t \ln x^t + y^t \ln y^t - (x^t + y^t) \ln(x^t + y^t) \right]. \tag{8.26}$$

Let $X = x^t$ and $Y = y^t$. Then the brackets in the expression (8.26) can be written as

$$x^t \ln x^t + y^t \ln y^t - (x^t + y^t) \ln(x^t + y^t) = \ln(X^X Y^Y) - \ln(X + Y)^{X+Y}. \tag{8.27}$$

Since $x, y > 0$ and $t > 0$, $X > 0$ and $Y > 0$. By these facts, we have

$$X^X < (X + Y)^X \quad \text{and} \quad Y^Y < (X + Y)^Y$$

and thus

$$X^X Y^Y < (X + Y)^X \times (X + Y)^Y = (X + Y)^{X+Y}. \tag{8.28}$$

Now by substituting the inequality (8.28) and the expression (8.27) into the right-hand side of the derivative (8.26), we obtain

$$f'(t) < 0$$

for all $t \in (0, \infty)$. By Theorem 8.9(b) and Remark 8.3, we see that $f(t)$ is strictly decreasing on $(0, \infty)$. Hence, if $\beta > \alpha > 0$, then we have

$$f(\alpha) > f(\beta)$$

which is exactly the inequality (8.25).   ∎

Problem 8.15

(★) *Prove that* $|\cos x - \cos y| \leq |x - y|$ *for every* $x, y \in \mathbb{R}$.

Proof. It is clear that the equality holds when $x = y$. Without loss of generality, we may assume that $x < y$. Since the function $f(t) = \cos t$ is continuous on $[x, y]$ and differentiable in (x, y), we obtain from the Mean Value Theorem for Derivatives that there exists a $p \in (x, y)$ such that

$$\cos y - \cos x = -(y - x) \sin p. \tag{8.29}$$

Since $|\sin t| \leq 1$ for all $t \in \mathbb{R}$, we know from this and the expression (8.29) that

$$|\cos x - \cos y| = |\sin p||x - y| \leq |x - y|.$$

This finishes the proof of the problem.   ∎

Remark 8.7

We say that a function $f : [a, b] \to \mathbb{R}$ satisfies a **Lipschitz condition** if there exists a positive constant $K > 0$ such that

$$|f(x) - f(y)| \leq K|x - y|$$

for all $x, y \in [a, b]$. Here the K is called a **Lipschitz constant**. Thus Problem 8.15 says that the function $f(x) = \cos x$ is Lipschitz. Furthermore, we can see from the proof of Problem 8.12 that a function whose derivative is bounded on (a, b) is also Lipschitz.

Problem 8.16

$(\star)(\star)$ *Suppose that $f : (a, b) \to \mathbb{R}$ is differentiable in (a, b) with unbounded derivative $f'(x)$. If f' is continuous on (a, b), prove that f is not Lipschitz on (a, b).*

Proof. Assume that f was Lipschitz on (a, b) with a Lipschitz constant K, i.e.,

$$|f(x) - f(y)| \leq K|x - y| \tag{8.30}$$

for all $x, y \in (a, b)$. Since f' is unbounded in (a, b), there exists a $p \in (a, b)$ such that

$$|f'(p)| > 2K.$$

Since $|f'| - 2K$ is continuous on (a, b) and (a, b) is open in $\mathbb{R}$, it yields from Problem 7.15 (The Sign-preserving Property) that there exists a $\delta > 0$ such that

$$|f'(x)| > 2K \tag{8.31}$$

for all $x \in (p - \delta, p + \delta) \subseteq (a, b)$. Pick $q \in (p - \delta, p)$. By the Mean Value Theorem for Derivatives, we know that

$$f(p) - f(q) = f'(\xi)(p - q) \tag{8.32}$$

for some $\xi \in (q, p)$. Combining the inequality (8.31), the expression (8.32) and then the Lipschitz condition (8.30), we have

$$2K|p - q| < |f'(\xi)| \times |p - q| = |f'(\xi)(p - q)| = |f(p) - f(q)| \leq K|p - q|$$

which implies $2K < K$, a contradiction. Hence f is not Lipschitz on (a, b) and we complete the proof of the problem. ∎

Problem 8.17

$(\star)(\star)$ *Suppose that $f : (1, +\infty) \to \mathbb{R}$ is differentiable in $(1, +\infty)$ and*

$$\lim_{x \to +\infty} f'(x) = 0.$$

Prove that

$$\lim_{x \to +\infty} \frac{f(x)}{x} = 0$$

holds.

Proof. By Definition 7.20 (Limits at Infinity), given $\epsilon > 0$, there exists a $M > 1$ such that $x > M$ implies that

$$|f'(x)| < \frac{\epsilon}{2}. \tag{8.33}$$

Pick $a > M$ and let $x > a$. Since f is differentiable in $(1, +\infty)$, it is also differentiable in (a, x) and continuous on $[a, x]$. By the Mean Value Theorem for Derivatives, we know that

$$f(x) - f(a) = f'(\xi)(x - a) \tag{8.34}$$

for some $\xi \in (a, x)$. Now we take absolute value to both sides of the expression (8.34) and then using the inequality (8.33), we get

$$|f(x) - f(a)| < \frac{\epsilon}{2}|x - a|.$$

Since $|f(x)| - |f(a)| \le |f(x) - f(a)|$, we have

$$|f(x)| \le |f(a)| + \frac{\epsilon}{2}|x - a|. \tag{8.35}$$

Since $f(a)$ is fixed, we may take $K > a$ large enough such that

$$\frac{|f(a)|}{K} < \frac{\epsilon}{2}. \tag{8.36}$$

Now if $x > K > a > M > 1$, then we follow from the inequalities (8.35) and (8.36) that

$$\left|\frac{f(x)}{x}\right| \le \left|\frac{f(a)}{x}\right| + \frac{\epsilon}{2} \cdot \frac{|x - a|}{|x|} < \frac{|f(a)|}{K} + \frac{\epsilon}{2} < \frac{\epsilon}{2} + \frac{\epsilon}{2} = \epsilon$$

which means that

$$\lim_{x \to +\infty} \frac{f(x)}{x} = 0$$

holds. We end the proof of the problem. $\blacksquare$

Problem 8.18

(⋆) *Find the number of solutions of the equation* $e^x = 1 - x$ *in* $\mathbb{R}$.

Proof. Let $f(x) = e^x + x - 1$. It is clear that

$$f(0) = e^0 + 0 - 1 = 1 + 0 - 1 = 0.$$

Thus 0 is a root of the equation $f(x) = 0$. Let $p \ne 0$ be another root of $f(x) = 0$. Without loss of generality, we may assume further that $p > 0$. Since f is differentiable in $(0, p)$, we apply the Mean Value Theorem for Derivatives to get

$$f(p) - f(0) = f'(\xi)(p - 0)$$

for some $\xi \in (0, p)$. Since $f(p) = 0$, this implies that

$$f'(\xi) = 0.$$

However, since $f'(x) = e^x + 1$, we have

$$e^\xi = -1$$

which is a contradiction. Hence we have finished the proof of the problem. $\blacksquare$

Problem 8.19

(⋆) (⋆) *Suppose that $a, b, c \in \mathbb{R}$ with $a < b < c$. Suppose further that $f : [a, c] \to \mathbb{R}$ is continuous on $[a, c]$ and differentiable in (a, c). Prove that there exist $\lambda \in (a, b)$ and $\theta \in (a, c)$ such that*

$$f(b) - f(a) = (b - a)f'(\lambda) \quad \text{and} \quad f(c) - f(a) = (c - a)f'(\theta),$$

where $\lambda < \theta$.

Proof. Since f is continuous on $[a, c]$ and differentiable in (a, c), it is also continuous on $[a, b]$ and $[b, c]$ as well as differentiable in (a, b) and (b, c). Applying the Mean Value Theorem for Derivatives twice, we know that there exist $\lambda \in (a, b)$ and $\xi \in (b, c)$ such that

$$f(b) - f(a) = (b - a)f'(\lambda) \quad \text{and} \quad f(c) - f(b) = (c - b)f'(\xi),$$

By these facts, it is obvious that

$$\begin{aligned}
\frac{f(c) - f(a)}{c - a} &= \frac{f(c) - f(b) + f(b) - f(a)}{c - a} \\
&= \frac{f(c) - f(b)}{c - b} \cdot \frac{c - b}{c - a} + \frac{f(b) - f(a)}{b - a} \cdot \frac{b - a}{c - a} \\
&= f'(\xi) \cdot \frac{c - b}{c - a} + f'(\lambda) \cdot \frac{b - a}{c - a}.
\end{aligned} \tag{8.37}$$

If we let $x = \frac{c-b}{c-a}$, then $1 - x = 1 - \frac{c-b}{c-a} = \frac{b-a}{c-a}$ so that the expression (8.37) can be rewritten as

$$\frac{f(c) - f(a)}{c - a} = xf'(\xi) + (1 - x)f'(\lambda). \tag{8.38}$$

Since $a < b < c$, we have $0 < x < 1$ and this implies that the value

$$xf'(\xi) + (1 - x)f'(\lambda)$$

lies between $f'(\xi)$ and $f'(\lambda)$. By the Intermediate Value Theorem for Derivatives and then using the expression (8.38), there exists a $\theta \in (\lambda, \xi)$ such that

$$f'(\theta) = xf'(\xi) + (1 - x)f'(\lambda) = \frac{f(c) - f(a)}{c - a},$$

i.e.,

$$f(c) - f(a) = (c - a)f'(\theta).$$

This completes the proof of the problem. ∎

Problem 8.20

(⋆) (⋆) *Let $p \in (a, b)$. Suppose that $f : [a, b] \to \mathbb{R}$ is continuous on $[a, b]$ and differentiable in $(a, b) \setminus \{p\}$. If $\lim\limits_{x \to p} f'(x)$ exists, prove that f is differentiable at p.*

Proof. We let

$$\lim_{x \to p} f'(x) = L$$

We want to show that

$$\lim_{x \to p} \frac{f(x) - f(p)}{x - p} = L \tag{8.39}$$

or equivalently, by Theorem 7.14,

$$F(p+) = F(p-) = L,$$

where

$$F(x) = \frac{f(x) - f(p)}{x - p}.$$

To this end, we pick a sequence $\{x_n\} \subseteq (p, b)$ converging to p. By Theorem 7.1, we have

$$\lim_{x \to p} f'(x) = \lim_{n \to \infty} f'(x_n) = L. \tag{8.40}$$

Now we consider the restrictions $f_{[p, x_n]} : [p, x_n] \to \mathbb{R}$. It is obvious that they are continuous on $[p, x_n]$ and differentiable in (p, x_n). By the Mean Value Theorem for Derivatives, there exist $\xi_n \in (p, x_n)$ such that

$$f(x_n) - f(p) = f'(\xi_n)(x_n - p). \tag{8.41}$$

Now $\xi_n \to p$ if and only if $x_n \to p$ and $x_n > p$. Therefore, we apply this fact and the limit (8.40) to get

$$L = \lim_{x \to p} f'(x) = \lim_{n \to \infty} f'(x_n) = \lim_{n \to \infty} f'(\xi_n). \tag{8.42}$$

By the expression (8.41) and the limit (8.42), we obtain

$$F(p+) = \lim_{n \to \infty} \frac{f(x_n) - f(p)}{x_n - p} = L.$$

Similarly, we can show that $F(p-) = L$. Hence we obtain the desired limit (8.39) so that f is differentiable at p, as required. This finishes the proof of the problem. ∎

Problem 8.21

(⋆)(⋆) *Suppose that $f : [a, b] \to \mathbb{R}$ is continuous on $[a, b]$ and differentiable in (a, b). Furthermore, let $f(a) = f(b) = 0$. Prove that for every $K \in \mathbb{R}$, there exists a $p \in (a, b)$ such that*

$$f'(p) = Kf(p).$$

Proof. Consider the function $F : [a, b] \to \mathbb{R}$ defined by

$$F(x) = e^{-Kx} f(x)$$

for every $x \in [a, b]$. It is clear that $F(a) = F(b) = 0$. By Theorem 8.7 (Rolle's Theorem), we see that there exists a point $p \in (a, b)$ such that

$$F'(p) = 0. \tag{8.43}$$

Since $F'(x) = e^{-Kx}[-Kf(x) + f'(x)]$ and $e^{-Kx} \neq 0$ on $[a, b]$, we follow from these and the result (8.43) that

$$f'(p) = Kf(p),$$

completing the proof of the problem.   ∎

Problem 8.22

$(\star)(\star)$ *Suppose that $f : [a, b] \to \mathbb{R}$ is continuous on $[a, b]$ and differentiable in (a, b). Prove that there exist $p, q \in (a, b)$ such that*

$$\frac{f'(p)}{a + b} = \frac{f'(q)}{2q}. \tag{8.44}$$

Proof. We apply the Mean Value Theorem for Derivatives to f, there exists a point $p \in (a, b)$ such that

$$f(b) - f(a) = (b - a)f'(p). \tag{8.45}$$

Next, we apply Theorem 8.8 (The Generalized Mean Value Theorem) to the functions $f(x)$ and $g(x) = x^2$ to conclude that there exists a $q \in (a, b)$ such that

$$2q[f(b) - f(a)] = (b^2 - a^2)f'(q) = (b - a)(b + a)f'(q). \tag{8.46}$$

Combining the expressions (8.45) and (8.46), we are able to show that

$$2q(b - a)f'(p) = (b - a)(b + a)f'(q)$$

which imply the formula (8.44) because $b - a \neq 0$. We complete the proof of the problem.   ∎

8.4 L'Hôspital's Rule

Problem 8.23

$(\star)(\star)$ *Reprove Problem 8.20 by using Theorem 8.10 (L'Hôspital's Rule).*

Proof. By Definition 8.1, we have to show that

$$\lim_{\substack{h \to 0 \\ h > 0}} \frac{f(p + h) - f(p)}{h} \quad \text{and} \quad \lim_{\substack{h \to 0 \\ h < 0}} \frac{f(p + h) - f(p)}{h}$$

exist and equal. Since f is continuous at p, we have

$$\lim_{\substack{h \to 0 \\ h > 0}} f(p + h) = \lim_{\substack{h \to 0 \\ h < 0}} f(p + h) = f(p)$$

by Theorem 7.14. By Theorem 8.10 (L'Hôspital's Rule), we follow that

$$\lim_{\substack{h \to 0 \\ h > 0}} \frac{f(p + h) - f(p)}{h} = \lim_{\substack{h \to 0 \\ h > 0}} \frac{f'(p + h)}{1} = \lim_{\substack{h \to 0 \\ h > 0}} f'(p + h) = f'(p+) \tag{8.47}$$

and

$$\lim_{\substack{h \to 0 \\ h < 0}} \frac{f(p+h) - f(p)}{h} = \lim_{\substack{h \to 0 \\ h < 0}} \frac{f'(p+h)}{1} = \lim_{\substack{h \to 0 \\ h < 0}} f'(p+h) = f'(p-). \tag{8.48}$$

Since $\lim\limits_{x \to p} f'(x)$ exists, we apply Theorem 7.14 to the limits (8.47) and (8.48) to conclude that

$$\lim_{\substack{h \to 0 \\ h > 0}} \frac{f(p+h) - f(p)}{h} = f'(p+) = f'(p-) = \lim_{\substack{h \to 0 \\ h < 0}} \frac{f(p+h) - f(p)}{h},$$

as desired. We end the proof of the problem. ∎

Problem 8.24

$(\star)$ Prove, by using Theorem 8.10 (L'Hôspital's Rule), that

$$\lim_{x \to 0} \frac{\sin x}{x} = 1.$$

Proof. If we let $f(x) = \sin x$ and $g(x) = x$, then it is easy to check that they satisfy the hypotheses of Theorem 8.10 (L'Hôspital's Rule). Since $f'(x) = \cos x$ and $g'(x) = 1$, we have

$$\lim_{x \to 0} \frac{\cos x}{1} = 1$$

which implies that our desired result, completing the proof of the problem. ∎

Problem 8.25

$(\star)(\star)$ Suppose that $f : (0, +\infty) \to \mathbb{R}$ is differentiable in $(0, +\infty)$ and $\lim\limits_{x \to +\infty} [f(x) + f'(x)] = L$. Prove that

$$\lim_{x \to +\infty} f(x) = L.$$

Proof. By the facts

$$f(x) = \frac{f(x)e^x}{e^x}$$

on $(0, +\infty)$, $(e^x)' \neq 0$ on $(0, +\infty)$ and $e^x \to +\infty$ as $x \to +\infty$, it follows from Theorem 8.11 that

$$\lim_{x \to +\infty} f(x) = \lim_{x \to +\infty} \frac{f(x)e^x}{e^x} = \lim_{x \to +\infty} \frac{[f(x) + f'(x)]e^x}{e^x} = \lim_{x \to +\infty} [f(x) + f'(x)] = L,$$

as desired. This finishes the proof of the problem. ∎

Problem 8.26

$(\star)(\star)$ Suppose that $f : (0, +\infty) \to \mathbb{R}$ is differentiable in $(0, +\infty)$ and $f'(x) + xf(x)$ is bounded on $(0, +\infty)$. Prove that

$$\lim_{x \to +\infty} f(x) = 0.$$

Proof. Since $f'(x) + xf(x)$ is bounded on $(0, +\infty)$, there exists a positive constant M such that

$$|f'(x) + xf(x)| \le M \tag{8.49}$$

for all $x \in (0, +\infty)$. By the facts

$$f(x) = \frac{f(x)e^{\frac{x^2}{2}}}{e^{\frac{x^2}{2}}}$$

on $(0, +\infty)$, $(e^{\frac{x^2}{2}})' = xe^{\frac{x^2}{2}} \ne 0$ on $(0, +\infty)$ and $e^{\frac{x^2}{2}} \to +\infty$ as $x \to +\infty$, it follows from Theorem 8.11 that

$$\lim_{x \to +\infty} f(x) = \lim_{x \to +\infty} \frac{f(x)e^{\frac{x^2}{2}}}{e^{\frac{x^2}{2}}} = \lim_{x \to +\infty} \frac{[f'(x) + xf(x)]e^{\frac{x^2}{2}}}{xe^{\frac{x^2}{2}}} = \lim_{x \to +\infty} \frac{f'(x) + xf(x)}{x}.$$

By using the bound (8.49) and Theorem 7.3 (Squeeze Theorem for Limits of Functions), we know that

$$\lim_{x \to +\infty} \frac{f'(x) + xf(x)}{x} = 0$$

which then implies our desired result that

$$\lim_{x \to +\infty} f(x) = 0,$$

completing the proof of the problem.   ■

Problem 8.27

$(\star)$ *Construct an example of functions f, g having a finite derivatives in $(0,1)$, $g'(x) \ne 0$ on $(0,1)$ and*

$$\lim_{\substack{x \to 0 \\ x > 0}} \frac{f(x)}{g(x)} = 0,$$

but $\displaystyle \lim_{\substack{x \to 0 \\ x > 0}} \frac{f'(x)}{g'(x)}$ *does not exist.*

Proof. Consider the functions $f(x) = \sin \frac{1}{x}$ and $g(x) = \frac{1}{x}$. It is clear that $g'(x) = -\frac{1}{x^2} \ne 0$ on $(0, 1)$. By direct computation, we have

$$\lim_{\substack{x \to 0 \\ x > 0}} \frac{f(x)}{g(x)} = \lim_{\substack{x \to 0 \\ x > 0}} \frac{\sin \frac{1}{x}}{\frac{1}{x}} = \lim_{\substack{x \to 0 \\ x > 0}} x \sin \frac{1}{x} = 0,$$

but

$$\lim_{\substack{x \to 0 \\ x > 0}} \frac{f'(x)}{g'(x)} = \lim_{\substack{x \to 0 \\ x > 0}} \frac{-x^{-2} \cos \frac{1}{x}}{-x^{-2}} = \lim_{\substack{x \to 0 \\ x > 0}} \cos \frac{1}{x}$$

which does not exist. We end the proof of the problem.   ■

Remark 8.8

Problem 8.27 shows that the conditions

$$\lim_{\substack{x \to a \\ x > a}} f(x) = \lim_{\substack{x \to a \\ x > a}} g(x) = 0 \quad \text{and} \quad \lim_{\substack{x \to a \\ x > a}} g(x) = \pm\infty$$

in Theorem 8.10 (L'Hôspital's Rule) and Theorem 8.11 cannot be omitted respectively.

8.5 Higher Order Derivatives and Taylor's Theorem

Problem 8.28

⊛ *Suppose that* $f : \mathbb{R} \to \mathbb{R}$ *is twice differentiable in* $\mathbb{R}$, $f(0) = 0$, $f(\frac{1}{2}) = \frac{1}{2}$ *and* $f'(0) = 0$. *Prove that*

$$f''(\xi) = 4$$

for some $\xi \in (0, \frac{1}{2})$.

Proof. Take $n = 2$, $x = \frac{1}{2}$ and $p = 0$ in Taylor's Theorem and Remark 8.5, we have

$$f\left(\frac{1}{2}\right) = f(0) + \frac{f'(0)}{2} + \frac{f''(\xi)}{8} \tag{8.50}$$

for some $\xi \in (0, \frac{1}{2})$. By the hypotheses, we obtain from the expression (8.50) that

$$\frac{1}{2} = \frac{f''(\xi)}{8}$$

which means that $f''(\xi) = 4$, completing the proof of the problem. ∎

Problem 8.29

⋆ *Prove Leibniz's Rule.*

Proof. When $n = 1$, the formula is clear because

$$(fg)' = f'g + g'f.$$

Assume that the statement is true for $n = m$ for some positive integer m, i.e.,

$$(fg)^{(m)} = \sum_{k=0}^{m} C_k^m f^{(m-k)} g^{(k)}.$$

When $n = k + 1$, we have

$$(fg)^{(m+1)} = \frac{\mathrm{d}}{\mathrm{d}x}\left(\sum_{k=0}^{m} C_k^m f^{(m-k)} g^{(k)}\right)$$

$$= \sum_{k=0}^{m} C_k^m \frac{\mathrm{d}}{\mathrm{d}x}\left(f^{(m-k)} g^{(k)}\right)$$

$$= \sum_{k=0}^{m} C_k^m [f^{(m-k+1)} g^{(k)} + f^{(m-k)} g^{(k+1)}]$$

$$= \sum_{k=0}^{m} C_k^m f^{(m-k+1)} g^{(k)} + \sum_{k=0}^{m} C_k^m f^{(m-k)} g^{(k+1)}$$

$$= \sum_{k=1}^{m} C_k^m f^{(m-k+1)} g^{(k)} + f^{(m+1)} g^{(0)} + \sum_{k=0}^{m-1} C_k^m f^{(m-k)} g^{(k+1)} + f^{(0)} g^{(m+1)}$$

$$= \sum_{k=0}^{m-1} C_{k+1}^m f^{(m-k)} g^{(k+1)} + f^{(m+1)} g^{(0)} + \sum_{k=0}^{m-1} C_k^m f^{(m-k)} g^{(k+1)} + f^{(0)} g^{(m+1)}$$

$$= \sum_{k=0}^{m-1} [C_{k+1}^m + C_k^m] f^{(m-k)} g^{(k+1)} + f^{(m+1)} g^{(0)} + f^{(0)} g^{(m+1)}$$

$$= \sum_{k=0}^{m-1} C_{k+1}^{m+1} f^{(m-k)} g^{(k+1)} + f^{(m+1)} g^{(0)} + f^{(0)} g^{(m+1)}$$

$$= \sum_{k=1}^{m} C_k^{m+1} f^{(m-k+1)} g^{(k)} + f^{(m+1)} g^{(0)} + f^{(0)} g^{(m+1)}$$

$$= \sum_{k=0}^{m+1} C_k^{m+1} f^{(m-k+1)} g^{(k)}.$$

Thus the statement is true for $n = m + 1$ if it is also true for $n = m$. Hence we follow from induction that Leibniz's Rule holds for all positive integers n. ■

Problem 8.30

$(\star)(\star)$ *Suppose that $f : [0,1] \to \mathbb{R}$ is a function such that f' is continuous on $[0,1]$ and f'' exists in $(0,1)$. Suppose that $f(0) = f(1)$ and there exists a $M > 0$ such that $|f''(x)| \leq M$ on $(0,1)$. Prove that*

$$|f'(x)| \leq \frac{M}{2}$$

for all $x \in (0,1)$.

Proof. Let $p \in (0,1)$. By Taylor's Theorem, there exists a ξ between x and p such that

$$f(x) = f(p) + f'(p)(x - p) + \frac{f''(\xi)}{2}(x - p)^2. \tag{8.51}$$

Putting $x = 0$ and $x = 1$ into the formula (8.51), we obtain

$$f(0) = f(p) - f'(p)p + \frac{f''(\xi_0)}{2}p^2 \tag{8.52}$$

and

$$f(1) = f(p) + f'(p)(1 - p) + \frac{f''(\xi_1)}{2}(1 - p)^2 \tag{8.53}$$

respectively, where $\xi_0 \in (0, p)$ and $\xi_1 \in (p, 1)$. By the hypothesis $f(0) = f(1)$ and considering the substraction of the formulas (8.52) and (8.53), we yield that

$$0 = f'(p) + \frac{f''(\xi_1)}{2}(1 - p)^2 - \frac{f''(\xi_0)}{2}p^2$$

and then

$$f'(p) = \frac{f''(\xi_0)}{2}p^2 - \frac{f''(\xi_1)}{2}(1 - p)^2. \tag{8.54}$$

Since $|f''(x)| \le M$ on $(0, 1)$, we get from the expression (8.54) that

$$|f'(p)| \le \frac{M}{2}[p^2 + (1 - p)^2]. \tag{8.55}$$

It is obvious that

$$p^2 + (1 - p)^2 = 2\left(p - \frac{1}{2}\right)^2 + \frac{1}{2} \le 1$$

for every $p \in (0, 1)$, we deduce from the inequality (8.55) that

$$|f'(p)| \le \frac{M}{2}$$

for all $p \in (0, 1)$. This completes the proof of the problem. ∎

Problem 8.31

$(\star)$ *Prove that for $x > 0$, we have*

$$\left| \ln(1 + x) - \left(x - \frac{x^2}{2} + \frac{x^3}{3} \right) \right| \le \frac{x^4}{4}. \tag{8.56}$$

Proof. Let $f(x) = \ln x$. Note that

$$f'(x) = \frac{1}{x}, \quad f''(x) = -\frac{1}{x^2}, \quad f'''(x) = \frac{2}{x^3} \quad \text{and} \quad f^{(4)}(x) = -\frac{6}{x^4}, \tag{8.57}$$

where $x > 0$. Applying Taylor's Theorem with $n = 4$, x replaced by $1 + x$ and $p = 1$, we have

$$f(1 + x) = f(1) + \frac{f'(1)}{1!}x + \frac{f''(1)}{2!}x^2 + \frac{f'''(1)}{3!}x^3 + \frac{f^{(4)}(\xi)}{4!}x^4 \tag{8.58}$$

for some $\xi \in (1, 1 + x)$. Using the expressions (8.57), we deduce from the formula (8.58) that

$$\ln(1 + x) = 0 + x - \frac{x^2}{2} + \frac{x^3}{3} - \frac{x^4}{4\xi^4}$$

so that

$$\left| \ln(1 + x) - \left(x - \frac{x^2}{2} + \frac{x^3}{3} \right) \right| = \frac{x^4}{4\xi^4}. \tag{8.59}$$

Since $\xi \ge 1$, our desired inequality (8.56) follows immediately from the expression (8.59). This completes the proof of the problem. ∎

> **Problem 8.32**
>
> ⊛ ⊛ *Suppose that* $f : [a, b] \to \mathbb{R}$ *is twice differentiable and* $f'(a) = f'(b) = 0$. *Prove that there exist* $\xi_1, \xi_2 \in (a, b)$ *such that*
> $$f(b) - f(a) = \frac{1}{2} \times \left(\frac{b-a}{2}\right)^2 [f''(\xi_1) - f''(\xi_2)].$$

Proof. By Taylor's Theorem with $n = 2$, $x = \frac{a+b}{2}$ and $p = a$, we have

$$f\left(\frac{a+b}{2}\right) = f(a) + f'(a)\left(\frac{b-a}{2}\right) + \frac{f''(\xi_1)}{2}\left(\frac{b-a}{2}\right)^2 \tag{8.60}$$

for some $\xi_1 \in (a, \frac{a+b}{2})$. Similarly, we apply Taylor's Theorem with $n = 2$, $x = \frac{a+b}{2}$ and $p = b$ to get

$$f\left(\frac{a+b}{2}\right) = f(b) - f'(b)\left(\frac{b-a}{2}\right) + \frac{f''(\xi_2)}{2}\left(\frac{b-a}{2}\right)^2 \tag{8.61}$$

for some $\xi_2 \in (\frac{a+b}{2}, b)$. Since $f'(a) = f'(b) = 0$, it follows from the subtraction of the expressions (8.60) and (8.61) that

$$f(b) - f(a) = \frac{1}{2} \times \left(\frac{b-a}{2}\right)^2 [f''(\xi_1) - f''(\xi_2)],$$

finishing the proof of the problem. ∎

> **Problem 8.33**
>
> ⊛ ⊛ ⊛ *Suppose that* $f : \mathbb{R} \to \mathbb{R}$ *is a twice differentiable function such that* $f''(x) \le 0$ *on* $\mathbb{R}$. *Prove that* f *is a constant function if* f *is bounded.*

Proof. Assume that $f'(p) > 0$ for some $p \in \mathbb{R}$. Take $x > p$. Then Taylor's Theorem shows that there exists a $\xi \in (p, x)$ such that

$$f(x) = f(p) + f'(p)(x - p) + \frac{(x - p)^2}{2} f''(\xi). \tag{8.62}$$

Since $f''(x) \le 0$ on $\mathbb{R}$, we know from the expression (8.62) that

$$f(x) \ge f(p) + f'(p)(x - p). \tag{8.63}$$

Since $f'(p) > 0$, it follows from the inequality (8.63) that

$$\lim_{x \to +\infty} f(x) = +\infty,$$

a contradiction. Thus we must have
$$f'(x) \le 0$$

for all $x \in \mathbb{R}$.

Next, we assume that $f'(p) < 0$ for some $p \in \mathbb{R}$. Take $x < p$ and we apply Taylor's Theorem again to ensure the existence of a $\theta \in (x, p)$ such that

$$f(x) = f(p) + f'(p)(x - p) + \frac{(x - p)^2}{2} f''(\theta). \tag{8.64}$$

Since $f''(x) \leq 0$ on $\mathbb{R}$, we derive from the expression (8.64) that

$$f(x) \geq f(p) + f'(p)(x - p). \tag{8.65}$$

We note that $x - p < 0$ so that the inequality (8.65) implies that

$$\lim_{x \to -\infty} f(x) = +\infty,$$

a contradiction again. Thus we have $f'(x) \geq 0$ for all $x \in \mathbb{R}$ and hence

$$f'(x) = 0$$

on $\mathbb{R}$. By Theorem 8.9(c), we conclude that f is a constant function, completing the proof of the problem. ∎

8.6 Convexity and Derivatives

Problem 8.34

$\bigstar$ $\bigstar$ Suppose that f is a real function defined in (a, b) and

$$\begin{vmatrix} 1 & 1 & 1 \\ x & y & z \\ f(x) & f(y) & f(z) \end{vmatrix} \geq 0$$

for every $x, y, z \in (a, b)$ and $x < y < z$. Prove that f is convex on (a, b).

Proof. Fix $x, y \in (a, b)$ and take $t \in (0, 1)$. Since $x < tx + (1 - t)y < y$, we have

$$\begin{vmatrix} 1 & 1 & 1 \\ x & tx + (1 - t)y & y \\ f(x) & f(tx + (1 - t)y) & f(y) \end{vmatrix} \geq 0. \tag{8.66}$$

By using properties of determinants, the determinant on the left-hand side in (8.66) becomes

$$\begin{vmatrix} 1 & 0 & 1 \\ x & 0 & y \\ f(x) & f(tx + (1 - t)y) - tf(x) - (1 - t)f(y) & f(y) \end{vmatrix}$$

$$= -[f(tx + (1 - t)y) - tf(x) - (1 - t)f(y)] \begin{vmatrix} 1 & 1 \\ x & y \end{vmatrix}$$

$$= -(y - x)[f(tx + (1 - t)y) - tf(x) - (1 - t)f(y)]. \tag{8.67}$$

By combining the inequality (8.66), the expression (8.67) and using the fact that $x < y$, we obtain

$$f(tx + (1 - t)y) - tf(x) - (1 - t)f(y) \leq 0$$

and this is equivalent to

$$f(tx + (1 - t)y) \leq tf(x) + (1 - t)f(y) \tag{8.68}$$

for all $t \in (0, 1)$. Since the equality in (8.68) holds trivially when $t = 0$ or $t = 1$, by Definition 8.13 (Convex Functions), f is convex on (a, b). This ends the proof of the problem. ∎

Problem 8.35

(⋆) *Prove that the function $f : (0, \pi) \to \mathbb{R}$ defined by*

$$f(x) = \frac{1}{\sin \frac{x}{2}}$$

is convex on $(0, \pi)$.

Proof. By direct differentiation, we have

$$f'(x) = -\frac{\cos \frac{x}{2}}{2 \sin^2 \frac{x}{2}} \quad \text{and} \quad f''(x) = \frac{1 + \cos^2 \frac{x}{2}}{4 \sin^3 \frac{x}{2}}.$$

For every $x \in (0, \pi)$, $\sin \frac{x}{2} > 0$ and $\cos \frac{x}{2} > 0$. Thus we have

$$f''(x) > 0$$

for all $x \in (0, \pi)$. By Theorem 8.16, f is (strictly) convex on $(0, \pi)$. This ends the proof of the problem. ∎

Problem 8.36

(⋆) (⋆) *Suppose that $f : (a, b) \to \mathbb{R}$ is a convex function. Let $x_1, x_2, \ldots, x_n \in (a, b)$ and $\alpha_1, \alpha_2, \ldots, \alpha_n$ are nonnegative constants such that $\alpha_1 + \alpha_2 + \cdots + \alpha_n = 1$. Prove that*

$$f(\alpha_1 x_1 + \cdots + \alpha_n x_n) \leq \alpha_1 f(x_1) + \cdots + \alpha_n f(x_n). \tag{8.69}$$

Proof. Since $\alpha_1, \alpha_2, \ldots, \alpha_n$ are nonnegative constants such that $\alpha_1 + \alpha_2 + \cdots + \alpha_n = 1$, we have

$$\alpha_1 x_1 + \cdots + \alpha_n x_n \in (a, b).$$

It is clear that the inequality (8.69) is true for $n = 1$. Assume that the statement is true for $n = m$ for some positive integer m, i.e.,

$$f(\alpha_1 x_1 + \cdots + \alpha_m x_m) \leq \alpha_1 f(x_1) + \cdots + \alpha_m f(x_m). \tag{8.70}$$

For $n = m + 1$, suppose that $x_1, \ldots, x_m, x_{m+1} \in (a, b)$ and $\alpha_1, \ldots, \alpha_m, \alpha_{m+1}$ are nonnegative constants such that $\alpha_1 + \cdots + \alpha_m + \alpha_{m+1} = 1$. Without loss of generality, we may assume that $0 < \alpha_{m+1} < 1$ and

$$\beta = \alpha_2 + \cdots + \alpha_{m+1} > 0.$$

Now it is evident that

$$\frac{\alpha_2}{\beta} + \cdots + \frac{\alpha_{m+1}}{\beta} = 1, \quad \alpha_1 + \beta = 1 \quad \text{and} \quad \frac{\alpha_2}{\beta} x_2 + \cdots + \frac{\alpha_{m+1}}{\beta} x_{m+1} \in (a, b).$$

Since f is convex on (a, b), we have

$$f(\alpha_1 x_1 + \cdots + \alpha_m x_m + \alpha_{m+1} x_{m+1}) = f\left(\alpha_1 x_1 + \beta\left(\frac{\alpha_2}{\beta} x_2 + \cdots + \frac{\alpha_{m+1}}{\beta} x_{m+1}\right)\right)$$

$$\leq \alpha_1 f(x_1) + \beta f\left(\frac{\alpha_2}{\beta} x_2 + \cdots + \frac{\alpha_{m+1}}{\beta} x_{m+1}\right). \quad (8.71)$$

Applying the assumption (8.70) to the second term of the right-hand side of the inequality (8.71), we obtain

$$f\left(\frac{\alpha_2}{\beta} x_2 + \cdots + \frac{\alpha_{m+1}}{\beta} x_{m+1}\right) \leq \frac{\alpha_2}{\beta} f(x_2) + \cdots + \frac{\alpha_{m+1}}{\beta} f(x_{m+1}). \quad (8.72)$$

By putting the inequality (8.72) back into the inequality (8.71), we conclude that

$$f(\alpha_1 x_1 + \cdots + \alpha_m x_m + \alpha_{m+1} x_{m+1}) \leq \alpha_1 f(x_1) + \beta\left[\frac{\alpha_2}{\beta} f(x_2) + \cdots + \frac{\alpha_{m+1}}{\beta} f(x_{m+1})\right]$$

$$= \alpha_1 f(x_1) + \cdots + \alpha_m f(x_m) + \alpha_{m+1} f(x_{m+1}).$$

Thus the statement is true for $n = m + 1$ if it is true for $n = m$. Hence it follows from induction that the inequality (8.69) holds for all positive integers n. We have completed the proof of the problem. ∎

Remark 8.9

The inequality (8.69) is called **Jensen's inequality**.

Problem 8.37

$(\star)(\star)$ *Suppose that $f(x) = -\ln x$. Prove that*

$$x_1^{\alpha_1} \cdots x_n^{\alpha_n} \leq \alpha_1 x_1 + \cdots + \alpha_n x_n$$

for all $x_1, \ldots, x_n \geq 0$ and $\alpha_1, \ldots, \alpha_n \geq 0$ with $\alpha_1 + \cdots + \alpha_n = 1$.

Proof. The inequality holds trivially when one of $x_1, \ldots, x_n$ is zero. Without loss of generality, we may assume that $x_1, \ldots, x_n > 0$. Since $f''(x) = \frac{1}{x^2} > 0$ for all $x > 0$, we deduce from Theorem 8.16 that f is convex on $(0, +\infty)$. By Problem 8.36, we have

$$-\ln(\alpha_1 x_1 + \cdots + \alpha_n x_n) \leq -\alpha_1 \ln x_1 - \cdots - \alpha_n \ln x_n$$

which implies that

$$x_1^{\alpha_1} \cdots x_n^{\alpha_n} \leq \alpha_1 x_1 + \cdots + \alpha_n x_n$$

as desired. This completes the proof of the problem. ∎

CHAPTER 9

The Riemann-Stieltjes Integral

There are two components in single variable calculus. One is differentiation which is reviewed in the previous chapter. The other one is **integration** which is the emphasis in this chapter. More precisely, we study properties of **definite integrals** (the integral of a real-valued function on a bounded interval) and the connection between derivatives and integrals. The main references for this part are [2, Chap. 1, 2 & 5], [3, Chap. 7], [5, Chap. 7], [6, Chap. 6], [13, Chap. 6] and [15, Chap. 11].

9.1 Fundamental Concepts

9.1.1 Definitions and Notations

A **partition** P of $[a, b]$ is a *finite* set of points, namely

$$P = \{x_0, x_1, \ldots, x_n\},$$

where $a = x_0 < x_1 < \cdots < x_{n-1} < x_n = b$. A partition P^* is called a **refinement** of P if $P \subseteq P^*$, i.e., every point of P is a point of P^*. If $\alpha : [a, b] \to \mathbb{R}$ is monotonically increasing, then we define, for each $k = 1, 2, \ldots, n$,

$$\Delta \alpha_k = \alpha(x_k) - \alpha(x_{k-1}) \tag{9.1}$$

so that

$$\sum_{k=1}^{n} \Delta \alpha_k = \alpha(b) - \alpha(a).$$

Since α is monotonically increasing, it is trivial from the definition (9.1) that $\Delta \alpha_k \geq 0$ for each $k = 1, 2, \ldots, n$.

Let $f : [a, b] \to \mathbb{R}$ be bounded and $I_k = [x_{k-1}, x_k]$, where $k = 1, 2, \ldots, n$. Now for each $k = 1, 2, \ldots, n$, we define

$$M_k = \sup_{x \in I_k} f(x) \quad \text{and} \quad m_k = \inf_{x \in I_k} f(x). \tag{9.2}$$

Definition 9.1 (The Riemann-Stieltjes Integral). *Suppose that $f : [a, b] \to \mathbb{R}$ is bounded and P is a partition of $[a, b]$. Furthermore, suppose that*

$$U(P, f, \alpha) = \sum_{k=1}^{n} M_k \Delta \alpha_k \quad and \quad L(P, f, \alpha) = \sum_{k=1}^{n} m_k \Delta \alpha_k,$$

where M_k and m_k are defined in (9.2). Finally, we define the two numbers

$$\overline{\int_a^b} f \, d\alpha = \inf_P U(P, f, \alpha) \quad and \quad \underline{\int_a^b} f \, d\alpha = \sup_P L(P, f, \alpha), \tag{9.3}$$

where the sup and inf in the numbers (9.3) take over all partitions P of $[a, b]$. If we have

$$\overline{\int_a^b} f \, d\alpha = \underline{\int_a^b} f \, d\alpha,$$

then we simply write the number as

$$\int_a^b f \, d\alpha \quad or \quad \int_a^b f(x) \, d\alpha(x) \tag{9.4}$$

*and we say that f is **integrable with respect to** α **in the Riemann sense** and write "$f \in \mathscr{R}(\alpha)$ on $[a, b]$". In this case, we call the number (9.4) the **Riemann-Stieltjes integral** of f with respect to α on $[a, b]$.*

Particularly, if we take $\alpha(x) = x$, then we write $U(P, f)$, $L(P, f)$ and $f \in \mathscr{R}$ instead of $U(P, f, \alpha)$, $L(P, f, \alpha)$ and $f \in \mathscr{R}(\alpha)$ respectively. In this case, the numbers (9.3)

$$\overline{\int_a^b} f \, dx = \inf_P U(P, f) \quad and \quad \underline{\int_a^b} f \, dx = \sup_P L(P, f)$$

are said to be the **upper Riemann integral** and the **lower Riemann integral** of f on $[a, b]$ respectively. Furthermore, the integral (9.4) is then called the **Riemann integral** and is denoted by

$$\int_a^b f(x) \, dx.$$

Remark 9.1

The number (9.4) depends **only** on a, b, f and α, but *not* the variable of integration x. In fact, the symbol x is a "dummy variable" and may be replaced by other symbols.

Two particular examples should be mentioned. It can be easily seen from Definition 9.1 (The Riemann-Stieltjes Integral) that

$$\int_a^b 1 \, d\alpha = \alpha(b) - \alpha(a) \quad and \quad \int_a^b 0 \, d\alpha = 0. \tag{9.5}$$

▌9.1.2 Criteria for Integrability of Real Functions and their Properties

Theorem 9.2 (The Riemann Integrability Condition). *We have $f \in \mathscr{R}(\alpha)$ on $[a, b]$ if and only if for every $\epsilon > 0$, there exists a partition P such that*

$$U(P, f, \alpha) - L(P, f, \alpha) < \epsilon. \tag{9.6}$$

In addition, the inequality (9.6) holds for every refinement of P.

Although the above result provides us a way to test whether f is integrable with respect to α on $[a, b]$ or not by simply checking the inequality (9.5), it is desirable to know some sufficient conditions for $f \in \mathscr{R}(\alpha)$ on $[a, b]$ and the following result serves this purpose.

Theorem 9.3. *Suppose that $f : [a, b] \to \mathbb{R}$ is bounded and $\alpha : [a, b] \to \mathbb{R}$ is monotonically increasing.*

(a) *If f is continuous on $[a, b]$, then we have $f \in \mathscr{R}(\alpha)$ on $[a, b]$.*

(b) *If f is monotonic on $[a, b]$ and α is continuous on $[a, b]$, then $f \in \mathscr{R}(\alpha)$ on $[a, b]$.*

(c) *If f has finitely many points of discontinuity on $[a, b]$ and α is continuous at every point at which f is discontinuous, then we have $f \in \mathscr{R}(\alpha)$ on $[a, b]$.*

Recall from Theorem 7.19 (Froda's Theorem) that the set of discontinuities of a monotonic function is **at most countable**. Thus Theorem 9.3 suggests a connection of the integrability of f with respect to α and the cardinalities of the discontinuities of f and α exists. In fact, if we employ a concept of Lebesgue's measure, then we can obtain something stronger for Riemann integrals.

Theorem 9.4 (The Lebesgue's Integrability Condition). *We have $f \in \mathscr{R}$ on $[a, b]$ if and only if the set of discontinuous points of f in $[a, b]$ is of **measure zero**.*

For the definition of a set of measure zero, please refer to [13, §11.4, 11.11, pp. 302, 303, 309]. In particular, any countable set is of measure zero, so it follows from Theorem 9.4 (The Lebesgue's Integrability Condition) that if $f : [a, b] \to \mathbb{R}$ is a bounded function whose points of discontinuity form a countable set, then $f \in \mathscr{R}$ on $[a, b]$.

Theorem 9.5 (Composition Theorem). *Suppose that $m \leq f(x) \leq M$ on $[a, b]$ for some constants M and m, $f \in \mathscr{R}(\alpha)$ on $[a, b]$ and $g : [m, M] \to \mathbb{R}$ is continuous. Then we have*

$$h = g \circ f \in \mathscr{R}(\alpha)$$

on $[a, b]$.

Theorem 9.6 (Operations of Integrable Functions).

(a) *If $f, g \in \mathscr{R}(\alpha)$ on $[a, b]$ and $A, B \in \mathbb{R}$, then $Af + Bg \in \mathscr{R}(\alpha)$ on $[a, b]$ and*

$$\int_a^b (Af + Bg) \, d\alpha = A \int_a^b f \, d\alpha + B \int_a^b g \, d\alpha.$$

(b) If $f, g \in \mathscr{R}(\alpha)$ on $[a, b]$ and $f(x) \leq g(x)$ on $[a, b]$, then we have

$$\int_a^b f \, d\alpha \leq \int_a^b g \, d\alpha.$$

(c) If $f \in \mathscr{R}(\alpha)$ on $[a, b]$ and $a < c < b$, then we have $f \in \mathscr{R}(\alpha)$ on $[a, c]$ and $[c, b]$. Besides, we have

$$\int_a^b f \, d\alpha = \int_a^c f \, d\alpha + \int_c^b f \, d\alpha.$$

(d) If $f \in \mathscr{R}(\alpha)$ and $f \in \mathscr{R}(\beta)$ on $[a, b]$ and $A, B \in \mathbb{R}^+$, then we have $f \in \mathscr{R}(A\alpha + B\beta)$ and

$$\int_a^b f \, d(A\alpha + B\beta) = A \int_a^b f \, d\alpha + B \int_a^b f \, d\beta.$$

(e) If $f, g \in \mathscr{R}(\alpha)$ on $[a, b]$, then $fg \in \mathscr{R}(\alpha)$ on $[a, b]$.

(f) If $f \in \mathscr{R}(\alpha)$ on $[a, b]$, then $|f| \in \mathscr{R}(\alpha)$ on $[a, b]$ and

$$\left| \int_a^b f \, d\alpha \right| \leq \int_a^b |f| \, d\alpha.$$

Apart from showing $f \in \mathscr{R}(\alpha)$ on $[a, b]$, people are also interested in evaluating the exact value of the number (9.4). To this end, we need the concept of the **unit step function**[a] I whose definition is given by

$$I(x) = \begin{cases} 0, & \text{if } x \leq 0; \\ 1, & \text{if } x > 0. \end{cases}$$

Then we have the following result:

Theorem 9.7. *Suppose that $\{c_n\}$ is a sequence of nonnegative numbers and $\{s_n\}$ is a sequence of **distinct** points in (a, b). If $\sum\limits_{n=1}^{\infty} c_n$ converges, $f : [a, b] \to \mathbb{R}$ is continuous and*

$$\alpha(x) = \sum_{n=1}^{\infty} c_n I(x - s_n),$$

then we have

$$\int_a^b f \, d\alpha = \sum_{n=1}^{\infty} c_n f(s_n).$$

∎ 9.1.3 The Substitution Theorem and the Change of Variables Theorem

The Substitution Theorem. *Suppose that $\alpha : [a, b] \to \mathbb{R}$ is monotonically increasing and $\alpha' \in \mathscr{R}$ on $[a, b]$. Then $f \in \mathscr{R}(\alpha)$ on $[a, b]$ if and only if $f\alpha' \in \mathscr{R}$ on $[a, b]$. Furthermore, we have*

$$\int_a^b f \, d\alpha = \int_a^b f(x)\alpha'(x) \, dx.$$

---

[a] Or called Heaviside step function.

The Change of Variables Theorem. *Suppose that $\varphi : [A, B] \to [a, b]$ is a strictly increasing continuous onto function. Furthermore, $\alpha : [a, b] \to \mathbb{R}$ is monotonically increasing on $[a, b]$ and $f \in \mathscr{R}(\alpha)$ on $[a, b]$. We define $\beta, g : [A, B] \to \mathbb{R}$ by*

$$\beta = \alpha \circ \varphi \quad \text{and} \quad g = f \circ \varphi.$$

Then we have $g \in \mathscr{R}(\beta)$ and

$$\int_a^b f \, d\alpha = \int_A^B g \, d\beta.$$

Remark 9.2

Both the Substitution Theorem and the Change of Variables Theorem provide us some convenient methods to *evaluate* the integral

$$\int_a^b f \, d\alpha.$$

Furthermore, we note from Remark 7.7 that φ is actually one-to-one.

9.1.4 The Fundamental Theorem of Calculus

There is a close connection between the concepts of differentiation and integration. In fact, there are two important and useful results related to this connection and they are classically combined and called the **First Fundamental Theorem of Calculus** and the **Second Fundamental Theorem of Calculus**.

The First Fundamental Theorem of Calculus. *Let $f \in \mathscr{R}$ on $[a, b]$ and $a \leq x \leq b$. Define*

$$F(x) = \int_a^x f(t) \, dt.$$

Then $F : [a, b] \to \mathbb{R}$ is continuous. In addition, if f is continuous at $p \in [a, b]$, then F is differentiable at p and

$$F'(p) = f(p)$$

holds.

The Second Fundamental Theorem of Calculus. *Let $f \in \mathscr{R}$ on $[a, b]$. If $F : [a, b] \to \mathbb{R}$ is a differentiable function such that $F' = f$, then we have*

$$\int_a^b f(x) \, dx = F(b) - F(a).$$

Remark 9.3

An **antiderivative** or a **primitive function** of a function f is a differentiable function F such that $F' = f$. Then the First Fundamental Theorem of Calculus implies the existence of antiderivatives for continuous functions and the Second Fundamental Theorem of Calculus comes up with a practical way of evaluating the integral by using a antiderivative F of f explicitly.

As an immediate application of the Second Fundamental Theorem of Calculus, we have the following important and practical skill in integral calculus: the **Integration by Parts**.

The Integration by Parts. *Suppose that* $F, G : [a, b] \to \mathbb{R}$ *are differentiable functions. Furthermore, we suppose that* $F', G' \in \mathscr{R}$ *on* $[a, b]$*. Then the following formula holds*

$$\int_a^b F(x)G'(x) \, \mathrm{d}x = F(b)G(b) - F(a)G(a) - \int_a^b F'(x)G(x) \, \mathrm{d}x.$$

9.1.5 The Mean Value Theorems for Integrals

In §8.1.4, we discuss the Mean Value Theorem for Derivatives. In integral calculus, we also have mean value theorems and the key message of one of them is to guarantee the *existence* of a rectangle with the same area and width.

The First Mean Value Theorem for Integrals. *Suppose that* $f : [a, b] \to \mathbb{R}$ *is a continuous function. Then there exists a* $p \in (a, b)$ *such that*

$$\int_a^b f(x) \, \mathrm{d}x = f(p)(b - a).$$

The Second Mean Value Theorem for Integrals. *Suppose that* $f : [a, b] \to \mathbb{R}$ *is monotonic increasing. Let* $A \le f(a+)$ *and* $B \ge f(b-)$*. If* $g : [a, b] \to \mathbb{R}$ *is continuous on* $[a, b]$*, then there exists a* $p \in [a, b]$ *such that*

$$\int_a^b f(x)g(x) \, \mathrm{d}x = A \int_a^p g(x) \, \mathrm{d}x + B \int_p^b g(x) \, \mathrm{d}x.$$

In particular, if $f(x) \ge 0$ *on* $[a, b]$*, then we have*

$$\int_a^b f(x)g(x) \, \mathrm{d}x = B \int_p^b g(x) \, \mathrm{d}x$$

for some $p \in [a, b]$*.*

Remark 9.4

The particular case of the Second Mean Value Theorem for Integrals is also known as **Bonnet's Theorem**.

9.2 Integrability of Real Functions

Problem 9.1

⭐ If $\alpha \equiv 0$ on $[a, b]$ and $f \in \mathscr{R}(\alpha)$ on $[a, b]$, prove that

$$\int_a^b f \, \mathrm{d}\alpha = 0.$$

Proof. Since $\alpha \equiv 0$ on $[a, b]$, we have $\Delta\alpha_k = 0$ by the expression (9.1). By Definition 9.1 (The Riemann-Stieltjes Integral), we get

$$U(P, f, \alpha) = L(P, f, \alpha) = 0$$

so that

$$\overline{\int_a^b} f \, \mathrm{d}\alpha = \underline{\int_a^b} f \, \mathrm{d}\alpha = 0.$$

Hence, it follows from the expression (9.4) that

$$\int_a^b f \, \mathrm{d}\alpha = 0.$$

This completes the proof of the problem. ∎

Problem 9.2

(⋆) *Suppose that $\theta \in \mathbb{R}$ and $f_\theta : [a, b] \to \mathbb{R}$ is defined by*

$$f_\theta(x) = \begin{cases} 1, & \text{if } x \in (a, b]; \\ \theta, & \text{if } x = a. \end{cases}$$

Prove that $f_\theta \in \mathscr{R}$ on $[a, b]$ for all $\theta \in \mathbb{R}$.

Proof. If $\theta = 1$, then we have $f_1 \equiv 1$ on $[a, b]$. By Theorem 9.3(a), it is clear that $f_1 \in \mathscr{R}$ on $[a, b]$. If $\theta \neq 1$, then f has *only* one point of discontinuity at $x = a$. By Theorem 9.3(c), we see that $f_\theta \in \mathscr{R}$ on $[a, b]$ in this case. Hence we have $f_\theta \in \mathscr{R}$ on $[a, b]$ for all $\theta \in \mathbb{R}$, completing the proof of the problem. ∎

Problem 9.3

(⋆) *Suppose that for every monotonic decreasing function $f : [a, b] \to \mathbb{R}$, we have*

$$\int_a^b f \, \mathrm{d}\alpha = 0.$$

Prove that α is a constant function on $[a, b]$.

Proof. Suppose that $p \in (a, b)$ and define

$$f(x) = \begin{cases} 1, & \text{if } a \leq x \leq p; \\ 0, & \text{if } p < x \leq b. \end{cases}$$

Since f is monotonic decreasing on $[0, 1]$, it follows from the hypothesis that

$$\int_a^b f \, \mathrm{d}\alpha = 0. \tag{9.7}$$

By Theorem 9.6(c) (Operations of Integrable Functions) and the examples (9.5), the equation (9.7) implies that

$$\int_a^p f \, d\alpha + \int_p^b f \, d\alpha = 0$$
$$\alpha(p) - \alpha(a) + 0 = 0$$
$$\alpha(p) = \alpha(a) \tag{9.8}$$

for every $p \in (a, b)$. Besides, if we take $f(x) = 1$ for all $x \in [a, b]$, then our hypothesis and the examples (9.5) again show that

$$\alpha(b) - \alpha(a) = \int_a^b d\alpha = 0$$

or equivalently

$$\alpha(b) = \alpha(a). \tag{9.9}$$

Combining the expressions (9.8) and (9.9), we conclude that α is a constant function on $[a, b]$, finishing the proof of the problem. ∎

Problem 9.4

$(\star)$ Suppose that $\alpha : [a, b] \to \mathbb{R}$ is a monotonically increasing function and $\mathscr{R}_\alpha[a, b]$ is the set of all bounded functions which are Riemann-Stieltjes integrable with respect to α on $[a, b]$. Prove that $\mathscr{R}_\alpha[a, b]$ is a vector space.

Proof. Recall that a set V is a **vector space** if

$$x + y \in V \quad \text{and} \quad cx \in V$$

for every $x, y \in V$ and scalar c. Then our desired result follows immediately from Theorem 9.6 (Operations of Integrable Functions), completing the proof of the problem. ∎

Problem 9.5

$(\star)\,(\star)$ Let $f : [a, b] \to \mathbb{R}$ be a function and $p \in (a, b)$. Suppose that $\alpha : [a, b] \to \mathbb{R}$ is monotonically increasing. Suppose, further that, there exists a $\epsilon > 0$ such that for every $\delta > 0$, we have $x, y \in (p, p + \delta)$ such that

$$|f(x) - f(p)| \geq \epsilon \quad \text{and} \quad \alpha(y) - \alpha(p) \geq \epsilon. \tag{9.10}$$

Prove that $f \notin \mathscr{R}(\alpha)$ on $[a, b]$.

Proof. Without loss of generality, we may assume that $\epsilon = 1$ in the inequalities (9.10). Suppose that $P = \{x_0, x_1, \ldots, x_n\}$ is a partition of $[a, b]$ such that $p = x_{j-1}$ for some $j = 2, 3, \ldots, n$. Consider the difference

$$U(P, f, \alpha) - L(P, f, \alpha) = \sum_{k=1}^n (M_k - m_k) \Delta \alpha_k. \tag{9.11}$$

Since we always have $M_k - m_k \geq 0$ for all $k = 1, 2, \ldots, n$ and $p = x_{j-1}$, we deduce from the expression (9.11) that

$$U(P, f, \alpha) - L(P, f, \alpha) \geq (M_j - m_j) \Delta \alpha_j$$
$$= (M_j - m_j)[\alpha(x_j) - \alpha(p)]. \tag{9.12}$$

By the assumption (9.10), we may take $x_j = y$ so that

$$\alpha(x_j) - \alpha(p) \geq 1.$$

It is clear from the other assumption (9.10) that

$$M_j - m_j \geq |f(x) - f(p)| \geq 1$$

for every $x \in (p, x_j]$. In fact, the inequality $M_j - m_j \geq 1$ also holds on $[p, x_j]$, so we obtain from the inequality (9.12) that

$$U(P, f, \alpha) - L(P, f, \alpha) \geq 1. \tag{9.13}$$

Assume that $f \in \mathscr{R}(\alpha)$ on $[a, b]$. By Theorem 9.2 (The Riemann Integrability Condition), we must have

$$U(P, f, \alpha) - L(P, f, \alpha) < 1 \tag{9.14}$$

for some partition P of $[a, b]$. Since the inequality (9.14) also holds for any refinement P^* of P, we may assume that $p \in P$. However, the inequality (9.14) will contradict the inequality (9.13) in this case. Hence we have shown that $f \notin \mathscr{R}(\alpha)$ on $[a, b]$ which completes the proof of the problem. ∎

Remark 9.5

The conditions (9.10) mean that both f and α are discontinuous from the right at p. Similarly, we can show that $f \notin \mathscr{R}(\alpha)$ on $[a, b]$ if they are discontinuous from the left at p.

Problem 9.6

$(\star)$ *Recall the Dirichlet function* $D(x)$ *is given by*

$$D(x) = \begin{cases} 1, & \text{if } [a, b] \cap \mathbb{Q}; \\ 0, & \text{otherwise.} \end{cases} \tag{9.15}$$

Prove that $D(x) \notin \mathscr{R}$ *on* $[a, b]$.

Proof. Let $P = \{x_0, x_1, \ldots, x_n\}$ be a partition of $[a, b]$. By the definition (9.15), we know that

$$M_k = \sup_{x \in I_k} D(x) = 1 \quad \text{and} \quad m_k = \inf_{x \in I_k} D(x) = 0,$$

where $I_k = [x_{k-1}, x_k]$ and $k = 1, 2, \ldots, n$. Thus we obtain from Definition 9.1 (The Riemann-Stieltjes Integral) that

$$U(P, D(x)) = b - a \quad \text{and} \quad L(P, D(x)) = 0$$

which imply that

$$\overline{\int_a^b} D(x)\,\mathrm{d}x = b - a \neq 0 = \underline{\int_a^b} D(x)\,\mathrm{d}x.$$

Hence we obtain $D(x) \notin \mathscr{R}$ on $[a,b]$ and we finish the proof of the problem. ∎

Problem 9.7

(⋆) *Construct a function $f : [a,b] \to \mathbb{R}$ such that $|f| \in \mathscr{R}$ on $[a,b]$, but $f \notin \mathscr{R}$ on $[a,b]$.*

Proof. Consider the function $f : [a,b] \to \mathbb{R}$ defined by

$$f(x) = \begin{cases} 1, & \text{if } x \in [a,b] \cap \mathbb{Q}; \\ -1, & \text{otherwise.} \end{cases}$$

Since $|f| \equiv 1$ on $[a,b]$, we must have $|f| \in \mathscr{R}$ on $[a,b]$. Assume that $f \in \mathscr{R}$ on $[a,b]$. Since

$$D(x) = \frac{1}{2}[f(x) + 1],$$

Theorem 9.6(a) (Operations of Integrable Functions) shows that $D(x) \in \mathscr{R}$ on $[a,b]$ which contradicts Problem 9.6. ∎

Problem 9.8

(⋆) *Suppose that $f : [0,1] \to \mathbb{R}$ and $g : [0,1] \to [0,1]$ are functions such that $f, g \in \mathscr{R}$ on $[0,1]$. Prove or disprove $f \circ g \in \mathscr{R}$ on $[0,1]$.*

Proof. Define $f : [0,1] \to \mathbb{R}$ by

$$f(x) = \begin{cases} 1, & \text{if } x \in (0,1]; \\ 0, & \text{if } x = 0. \end{cases}$$

By Theorem 9.3(c), we have $f \in \mathscr{R}$ on $[0,1]$. By [13, Exercise 18, p. 100], we know that the function $g : [0,1] \to [0,1]$ defined by[b]

$$g(x) = \begin{cases} \frac{1}{n}, & \text{if } x = \frac{m}{n}, m \in \mathbb{Z}, n \in \mathbb{N}, m \text{ and } n \text{ are coprime, } x \in [0,1]; \\ 0, & \text{otherwise} \end{cases} \tag{9.16}$$

is continuous at every irrational point of $[0,1]$ and discontinuous at every rational point of $[0,1]$. We have to show that $g \in \mathscr{R}$ on $[0,1]$. Since the set of all rational points of $[0,1]$ is countable, it is of measure zero. By Theorem 9.4 (The Lebesgue's Integrability Condition), we see that $g \in \mathscr{R}$ on $[0,1]$. However, it is easy to check that

$$D(x) = f(g(x))$$

on $[0,1]$, where D is the Dirichlet function. By Problem 9.6, we know that $D(x) \notin \mathscr{R}$ on $[0,1]$. This ends the proof of the problem. ∎

---

[b]We notice that 1 is the only positive integer which is coprime to 0.

Remark 9.6

The function $g(x)$ defined in (9.16) is called the **Riemann function**, the **Thomae's function**, the **popcorn function** or the **ruler function**.

Problem 9.9

⊛ ⊛ *Suppose that $f : [a, b] \to \mathbb{R}$ is bounded and there exists a sequence $\{P_n\}$ of partitions of $[a, b]$ such that*

$$\lim_{n \to \infty} [U(P_n, f) - L(P_n, f)] = 0. \tag{9.17}$$

Prove that $f \in \mathscr{R}$ on $[a, b]$ and

$$\int_a^b f(x) \, dx = \lim_{n \to \infty} U(P_n, f) = \lim_{n \to \infty} L(P_n, f).$$

Proof. Given $\epsilon > 0$. By the hypothesis (9.17), there exists a positive integer N such that $n \geq N$ implies

$$U(P_n, f) - L(P_n, f) = |U(P_n, f) - L(P_n, f) - 0| < \epsilon. \tag{9.18}$$

Thus it follows from Theorem 9.2 (The Riemann Integrability Condition) that $f \in \mathscr{R}$ on $[a, b]$. For the second assertion, we obtain from the paragraph following Definition 9.1 (The Riemann-Stieltjes Integral) that

$$L(P_n, f) \leq \sup_P L(P, f) = \int_a^b f(x) \, dx = \inf_P U(P, f) \leq U(P_n, f).$$

Therefore, we establish from this and the inequality (9.18) that

$$\left| U(P_n, f) - \int_a^b f(x) \, dx \right| \leq |U(P_n, f) - L(P_n, f)| < \epsilon$$

for every $n \geq N$. Thus we have

$$\int_a^b f(x) \, dx = \lim_{n \to \infty} U(P_n, f). \tag{9.19}$$

Similarly, we can show that the expression (9.19) also holds when $U(P_n, f)$ is replaced by $L(P_n, f)$. This completes the proof of the problem. ∎

Problem 9.10

⊛ ⊛ *Suppose that $f, g \in \mathscr{R}(\alpha)$ on $[a, b]$. Prove the **Schwarz Inequality for Integral***

$$\left| \int_a^b f g \, d\alpha \right| \leq \left(\int_a^b f^2 \, d\alpha \right)^{\frac{1}{2}} \left(\int_a^b g^2 \, d\alpha \right)^{\frac{1}{2}}. \tag{9.20}$$

Proof. By Theorem 9.6(e) (Operations of Integrable Functions), we know that $f^2, g^2, fg \in \mathscr{R}(\alpha)$ on $[a, b]$. We consider the case that

$$\int_a^b f^2 \, d\alpha > 0 \quad \text{and} \quad \int_a^b g^2 \, d\alpha > 0.$$

Let

$$F(x) = \frac{f(x)}{\left(\int_a^b f^2 \, d\alpha \right)^{\frac{1}{2}}} \quad \text{and} \quad G(x) = \frac{g(x)}{\left(\int_a^b g^2 \, d\alpha \right)^{\frac{1}{2}}}. \tag{9.21}$$

Since $f^2, g^2, fg \in \mathscr{R}(\alpha)$ on $[a, b]$, it is obvious from Theorem 9.6 (Operations of Integrable Functions) that $F^2, G^2, FG \in \mathscr{R}(\alpha)$ on $[a, b]$ and

$$\int_a^b F^2 \, d\alpha = \int_a^b G^2 \, d\alpha = 1.$$

By applying the A.M. $\geq$ G.M. to F^2 and G^2, we see that

$$\int_a^b FG \, d\alpha \leq \int_a^b \left(\frac{F^2 + G^2}{2} \right) d\alpha = \frac{1}{2} \int_a^b F^2 \, d\alpha + \frac{1}{2} \int_a^b G^2 \, d\alpha = 1. \tag{9.22}$$

Thus, after putting the two expressions (9.21) into the inequality (9.22), we get

$$\int_a^b fg \, d\alpha \leq \left(\int_a^b f^2 \, d\alpha \right)^{\frac{1}{2}} \left(\int_a^b g^2 \, d\alpha \right)^{\frac{1}{2}}. \tag{9.23}$$

Next, we suppose that

$$\int_a^b f^2 \, d\alpha = 0.$$

Given $\epsilon > 0$. Then the A.M. $\geq$ G.M. implies that

$$fg = (\epsilon^{-1} f)(\epsilon g) \leq \frac{(\epsilon^{-1} f)^2 + (\epsilon g)^2}{2}.$$

By Theorem 9.6 (Operations of Integrable Functions), we see that

$$\int_a^b fg \, d\alpha \leq \frac{\epsilon^{-2}}{2} \int_a^b f^2 \, d\alpha + \frac{\epsilon^2}{2} \int_a^b g^2 \, d\alpha = \frac{\epsilon^2}{2} \int_a^b g^2 \, d\alpha. \tag{9.24}$$

Since ϵ is arbitrary, it follows from the inequality (9.24) that

$$\int_a^b fg \, d\alpha \leq 0. \tag{9.25}$$

By a similar argument, we can show that the inequality (9.25) also holds when

$$\int_a^b g^2 \, d\alpha = 0.$$

Hence what we have shown is that the inequality (9.23) holds *for any* $f, g \in \mathscr{R}$ on $[a, b]$.

Finally, if we replace f by $-f$ in the inequality (9.23), then we achieve

$$-\int_a^b fg \,\mathrm{d}\alpha \le \left(\int_a^b (-f)^2 \,\mathrm{d}\alpha\right)^{\frac{1}{2}} \left(\int_a^b g^2 \,\mathrm{d}\alpha\right)^{\frac{1}{2}}$$

$$= \left(\int_a^b f^2 \,\mathrm{d}\alpha\right)^{\frac{1}{2}} \left(\int_a^b g^2 \,\mathrm{d}\alpha\right)^{\frac{1}{2}}. \tag{9.26}$$

Hence the expected inequality (9.20) follows immediately from combining the inequalities (9.23) and (9.26). This completes the proof of the problem. ∎

Problem 9.11

(⋆) Given that $\displaystyle\int_0^1 x^2 \,\mathrm{d}x = \frac{1}{3}$. Suppose that $f \in \mathscr{R}$ on $[0,1]$ and

$$\int_0^1 f(x) \,\mathrm{d}x = \int_0^1 x f(x) \,\mathrm{d}x = 1.$$

Prove that

$$\int_0^1 f^2(x) \,\mathrm{d}x \ge 3. \tag{9.27}$$

Proof. By using Problem 9.10 directly with $g(x) = x$, we gain

$$\int_0^1 f^2(x) \,\mathrm{d}x \times \int_0^1 x^2 \,\mathrm{d}x \ge \left(\int_0^1 x f(x) \,\mathrm{d}x\right)^2 = 1. \tag{9.28}$$

By the given hypothesis

$$\int_0^1 x^2 \,\mathrm{d}x = \frac{1}{3},$$

the desired result (9.27) follows immediately from the inequality (9.28). Hence we have completed the proof of the problem. ∎

Problem 9.12

(⋆)(⋆) Suppose that $\alpha : [a,b] \to \mathbb{R}$ is a monotonically increasing function and $f : [a,b] \to \mathbb{R}$ is a bounded function such that $f \in \mathscr{R}(\alpha)$ on $[a,b]$. Recall from Problem 7.12 that

$$\varphi(x) = \max(f(x), 0) \quad \text{and} \quad \psi(x) = \min(f(x), 0)$$

Prove that $\varphi, \psi \in \mathscr{R}(\alpha)$ on $[a,b]$.

Proof. By the proof of Problem 7.12, we see that

$$\varphi(x) = \frac{1}{2}[f(x) + |f(x)|] \quad \text{and} \quad \psi(x) = \frac{1}{2}[f(x) - |f(x)|].$$

By Theorem 9.6(f) and then (a) (Operations of Integrable Functions), we are able to conclude that $\varphi, \psi \in \mathscr{R}(\alpha)$ on $[a,b]$. We have completed the proof of the problem. ∎

9.3 Applications of Integration Theorems

Problem 9.13

(⋆) Suppose that $f : [a, b] \to \mathbb{R}$ is bounded and $f \in \mathscr{R}$ on $[a, b]$ and $a \le x \le b$. Define

$$F(x) = \int_a^x f(t)\,\mathrm{d}t.$$

Prove that there exists a positive constant M such that

$$|F(x) - F(y)| \le M|x - y| \tag{9.29}$$

for all $x, y \in [a, b]$.

Proof. Since f is bounded on $[a, b]$, there exists a positive constant M such that

$$|f(x)| \le M \tag{9.30}$$

for all $x \in [a, b]$. Let $x \ge y$. Then we deduce from Theorem 9.6(c), (f) (Operations of Integrable Functions) and the bound (9.30) that

$$
\begin{aligned}
|F(x) - F(y)| &= \left| \int_a^x f(t)\,\mathrm{d}t - \int_a^y f(t)\,\mathrm{d}t \right| \\
&= \left| \int_a^y f(t)\,\mathrm{d}t + \int_y^x f(t)\,\mathrm{d}t - \int_a^y f(t)\,\mathrm{d}t \right| \\
&= \left| \int_y^x f(t)\,\mathrm{d}t \right| \\
&\le \left| \int_y^x |f(t)|\,\mathrm{d}t \right| \\
&\le \left| \int_y^x M\,\mathrm{d}t \right| \\
&= M|x - y|.
\end{aligned}
$$

The case for $x < y$ is similar. Hence the inequality (9.29) holds for all $x, y \in [a, b]$ and we have finished the proof of the problem. ∎

Problem 9.14

(⋆)(⋆) Suppose that $\alpha : [0, 4] \to \mathbb{R}$ is a function defined by

$$
\alpha(x) = \begin{cases} x^2, & \text{if } 0 \le x \le 2; \\[2mm] x^4, & \text{if } 2 < x \le 4. \end{cases}
$$

Evaluate the integral

$$\int_0^4 x\,\mathrm{d}\alpha. \tag{9.31}$$

Proof. It is clear that α is a monotonically increasing function and it has a discontinuity of the first kind at 2 (see Definition 7.15 (Types of Discontinuity)). By Theorem 9.3(a), the integral (9.31) is well-defined. Now we write $\alpha = \beta + \gamma$, where $\beta, \gamma : [0, 4] \to \mathbb{R}$ are functions defined by

$$\beta(x) = \begin{cases} 0, & \text{if } 0 \le x \le 2; \\ 12, & \text{if } 2 < x \le 4 \end{cases} \quad \text{and} \quad \gamma(x) = \begin{cases} x^2, & \text{if } 0 \le x \le 2; \\ x^4 - 12, & \text{if } 2 < x \le 4. \end{cases}$$

Then it is obvious that both β and γ are monotonically increasing functions on $[0, 4]$. By Theorem 9.6(d) and then (c)(Operations of Integrable Functions), we obtain

$$\int_0^4 x \, d\alpha = \int_0^4 x \, d\beta + \int_0^4 x \, d\gamma = \int_0^4 x \, d\beta + \int_0^2 x \, d\gamma + \int_2^4 x \, d\gamma. \qquad (9.32)$$

We note that $(x^2)' = 2x \in \mathscr{R}$ on $[0, 2]$ and $(x^4 - 12)' = 4x^3 \in \mathscr{R}$ on $[2, 4]$, so the Substitution Theorem yields

$$\int_0^2 x \, d\gamma = \int_0^2 2x^2 \, dx = \frac{16}{3}$$

and

$$\int_2^4 x \, d\gamma = \int_2^4 4x^4 \, dx = \left[\frac{4x^5}{5} \right]_2^4 = \frac{3968}{5},$$

so the integral (9.32) reduces to

$$\int_0^4 x \, d\alpha = \int_0^4 x \, d\beta + \frac{16}{3} + \frac{3968}{5}. \qquad (9.33)$$

Since $\beta(x) = 12I(x - 2)$, we follow from Theorem 9.7 that

$$\int_0^4 x \, d\beta = 12 \times 2 = 24. \qquad (9.34)$$

Hence we obtain from the integrals (9.33) and (9.34) that

$$\int_0^4 x \, d\alpha = 24 + \frac{16}{3} + \frac{3968}{5} = \frac{12344}{15},$$

completing the proof of the problem. ∎

Problem 9.15

$\bigstar$ $\bigstar$ $\bigstar$ *Suppose that $f \in \mathscr{R}$ on $[a, b]$ and $g : [a, b] \to \mathbb{R}$ is a function such that the set*

$$E = \{x \in [a, b] \mid f(x) \ne g(x)\}$$

is finite. Prove that $g \in \mathscr{R}$ on $[a, b]$ and

$$\int_a^b f(x) \, dx = \int_a^b g(x) \, dx$$

without using Theorem 9.4 (The Lebesgue's Integrability Condition).

Proof. Given $\epsilon > 0$. Let $h : [a, b] \to \mathbb{R}$ be defined by

$$h(x) = f(x) - g(x)$$

and $E = \{x_1, \ldots, x_n\}$ for some positive integer n. There are three cases for consideration.

- **Case (1):** $a < x_1 < \cdots < x_n < b$. Let

$$M = \max(|h(x_1)|, |h(x_2)|, \ldots, |h(x_n)|) > 0.$$

By Theorem 2.2 (Density of Rationals), there exists a rational $\delta > 0$ such that

$$a < x_1 - \delta, \quad x_n + \delta < b \quad \text{and} \quad x_k + \delta < x_{k+1} - \delta,$$

where $k = 1, 2, \ldots, n - 1$. Furthermore, we may assume that

$$\delta < \frac{\epsilon}{2Mn}. \tag{9.35}$$

We consider the partition

$$P = \{a, x_1 - \delta, x_1 + \delta, x_2 - \delta, x_2 + \delta, \ldots, x_n - \delta, x_n + \delta, b\}.$$

For each $k = 1, 2, \ldots, n$, in the interval $I_k = [x_k - \delta, x_k + \delta]$, since $|h(x)| \le M$ for all $x \in [a, b]$, we have[c]

$$-M \le M_k = \sup_{I_k} h(x) \le M \quad \text{and} \quad -M \le m_k = \inf_{I_k} h(x) \le M. \tag{9.36}$$

In the interval $I = [a, x_1 - \delta]$ or $I' = [x_n + \delta, b]$, we have $h(x) = 0$ so that

$$\begin{aligned}
M_I = \sup_I h(x) = 0, \quad m_I = \inf_I h(x) = 0, \\
M_{I'} = \sup_{I'} h(x) = 0, \quad m_{I'} = \inf_{I'} h(x) = 0.
\end{aligned} \tag{9.37}$$

By definition, the sup and the inf given in (9.37), we have

$$U(P, h) = M_I(x_j - \delta - a) + \sum_{k=1}^{n} M_k(2\delta) + M_{I'}(b - x_j - \delta) = 2\delta \sum_{k=1}^{n} M_k$$

and

$$L(P, h) = m_I(x_j - \delta - a) + \sum_{k=1}^{n} m_k(2\delta) + m_{I'}(b - x_j - \delta) = 2\delta \sum_{k=1}^{n} m_k.$$

Thus we deduce from the inequality (9.35), the sup and the inf in (9.36) that

$$-\epsilon < -2nM\delta \le U(P, h) \le 2nM\delta < \epsilon \tag{9.38}$$

and

$$-\epsilon < -2nM\delta \le L(P, h) \le 2nM\delta < \epsilon. \tag{9.39}$$

---

[c]It may happen that $h(x_k) < 0$ for some k.

Since ϵ is arbitrary, we get from the estimates (9.38) and (9.39) that

$$U(P, h) = L(P, h) = 0.$$

Now we are able to conclude from the paragraph following Definition 9.1 (The Riemann-Stieltjes Integral) that

$$\int_a^b h(x)\,\mathrm{d}x = \overline{\int_a^b} h(x)\,\mathrm{d}x = \underline{\int_a^b} h(x)\,\mathrm{d}x = 0. \tag{9.40}$$

Hence we apply Theorem 9.6(a) (Operations of Integrable Functions) to establish that

$$\int_a^b f(x)\,\mathrm{d}x = \int_a^b f(x)\,\mathrm{d}x - \int_a^b h(x)\,\mathrm{d}x = \int_a^b [f(x) - h(x)]\,\mathrm{d}x = \int_a^b g(x)\,\mathrm{d}x. \tag{9.41}$$

- **Case (2):** $a = x_1$. In this case, we still have

$$M = \max(|h(a)|, |h(x_2)|, \ldots, |h(x_n)|) > 0$$

and a $\delta > 0$ such that

$$a + \delta < x_2 - \delta, \quad x_n + \delta < b \quad \text{and} \quad x_k + \delta < x_{k+1} - \delta,$$

where $k = 2, 3, \ldots, n - 1$. By this setting, it is easy to check that the inequalities (9.36) also hold for $k = 2, 3, \ldots, n$ and

$$M_{I'} = m_{I'} = 0, \tag{9.42}$$

where $I' = [x_n + \delta, b]$. Now we have to check the remaining interval $I_1 = [a, a + \delta]$. Since $h(a) \neq 0$, it is evident that

$$-M \leq M_1 = \sup_{I_1} h(x) \leq M \quad \text{and} \quad -M \leq m_1 = \inf_{I_1} h(x) \leq M. \tag{9.43}$$

Thus we combine the inequalities (9.36), (9.43) and the values (9.42) to get the inequalities (9.38) and (9.39). Hence, by a similar argument as in the proof of **Case (1)**, we see that the integral (9.40) and then the integral equation (9.41) also hold in this case.

- **Case (3):** $b = x_n$. Since the argument of this part is very similar to that proven in **Case (2)**, we omit the details here.

We have completed the proof of the problem. ∎

Problem 9.16

$\star$ $\star$ $\star$ *Suppose that $\mathscr{C}([0, 1])$ denotes the set of all continuous real functions on $[0, 1]$. For every $f, g \in \mathscr{C}([0, 1])$, we define*

$$d(f, g) = \int_0^1 \frac{|f(x) - g(x)|}{1 + |f(x) - g(x)|}\,\mathrm{d}x. \tag{9.44}$$

Prove that d is a metric in $\mathscr{C}([0, 1])$. Is $\mathscr{C}([0, 1])$ a complete metric space with respect to this metric?

Proof. Since f and g are continuous functions on $[0,1]$, $f, g \in \mathscr{R}$ on $[0,1]$ and Theorem 9.6 (Operations of Integrable Functions) guarantees that the integral (9.44) is well-defined.

Next, we recall from Problem 2.7 that

$$\frac{|c|}{1+|c|} \le \frac{|a|}{1+|a|} + \frac{|b|}{1+|b|}, \tag{9.45}$$

where $c = a + b$. If we put $a = f - g$ and $b = g - h$, then $c = f - h$ and the inequality (9.45) implies that

$$\frac{|f-h|}{1+|f-h|} \le \frac{|f-g|}{1+|f-g|} + \frac{|g-h|}{1+|g-h|}. \tag{9.46}$$

Thus we obtain from the inequality (9.46) that

$$d(f,h) \le d(f,g) + d(g,h)$$

which proves the triangle inequality is valid. Hence d is a metric in $\mathscr{C}([0,1])$.

We claim that $\mathscr{C}([0,1])$ is *not* a complete metric space. To this end, we consider the functions $f_n : [0,1] \to \mathbb{R}$ defined by

$$f_n(x) = \begin{cases} n^2 x, & \text{if } 0 \le x < \frac{1}{n}; \\ \\ \dfrac{1}{x}, & \text{if } \frac{1}{n} \le x \le 1. \end{cases} \tag{9.47}$$

It is clear that each f_n is continuous at every point on $[0,1]$ *except possibly* the point $\frac{1}{n}$. We check the continuity of f_n at $\frac{1}{n}$. Since $f_n(\frac{1}{n}+) = \frac{1}{\frac{1}{n}} = n$ and $f_n(\frac{1}{n}-) = n^2 \times \frac{1}{n} = n$, each f_n is continuous at $\frac{1}{n}$. In other words, we have $\{f_n\} \subseteq \mathscr{C}([0,1])$.

Given $\epsilon > 0$. Take N to be a positive integer such that $\frac{1}{N} < \epsilon$. Now if $x \ge \max(\frac{1}{m}, \frac{1}{n})$, then we have $f_n(x) = f_m(x) = \frac{1}{x}$ so that

$$\int_{\max(\frac{1}{m},\frac{1}{n})}^{1} \frac{|f_n(x) - f_m(x)|}{1 + |f_n(x) - f_m(x)|} \, dx = 0. \tag{9.48}$$

Therefore, for $m, n \ge N$, we deduce from the definition (9.44) and the result (9.48) that

$$d(f_m, f_n) = \int_0^{\max(\frac{1}{m},\frac{1}{n})} \frac{|f_n(x) - f_m(x)|}{1 + |f_n(x) - f_m(x)|} \, dx \le \int_0^{\max(\frac{1}{m},\frac{1}{n})} dx = \max\left(\frac{1}{m}, \frac{1}{n}\right) \le \frac{1}{N} < \epsilon.$$

By Definition 5.12, $\{f_n\}$ is a Cauchy sequence.

Assume that $\mathscr{C}([0,1])$ was complete. It means that $\{f_n\}$ converges to a function $f \in \mathscr{C}([0,1])$, see the paragraph following Theorem 5.13. By the definition of f_n in (9.47), it is reasonable to *conjecture* that

$$f(x) = \frac{1}{x}$$

on $(0,1]$. In fact, if $f(p) \ne \frac{1}{p}$ *for some* $p \in (0,1]$, then the continuity of f ensures that there exist $\epsilon > 0$ and $\delta > 0$ such that

$$\left| f(x) - \frac{1}{x} \right| \ge \epsilon \tag{9.49}$$

for all $x \in [p - \delta, p]$. Without loss of generality, we may take $\epsilon = \delta$. Since the function $g(x) = f(x) - \frac{1}{x}$ is continuous on $[p - \epsilon, p]$, it follows from the Extreme Value Theorem that there exists a positive constant M such that

$$\left| f(x) - \frac{1}{x} \right| \leq M \tag{9.50}$$

on $[p - \delta, p]$. Now, for sufficiently large n, we have $[p - \epsilon, p] \subseteq (\frac{1}{n}, 1]$ so that $f_n(x) = \frac{1}{x}$. Thus we follow from this and the inequalities (9.49) and (9.50) that

$$d(f_n, f) = \int_0^1 \frac{|f(x) - f_n(x)|}{1 + |f(x) - f_n(x)|} \, \mathrm{d}x \geq \int_{p-\epsilon}^p \frac{|f(x) - \frac{1}{x}|}{1 + |f(x) - \frac{1}{x}|} \, \mathrm{d}x \geq \int_{p-\epsilon}^p \frac{\epsilon}{1 + M} \, \mathrm{d}x > 0.$$

In other words, f_n *does not* converge to f and then

$$f(p) = \frac{1}{p}$$

for all $p \in (0, 1]$, but this contradicts the continuity of f on $[0, 1]$. Hence $\mathscr{C}([0, 1])$ is not complete with respect to the metric d and we have completed the proof of the problem. ∎

Problem 9.17

⋆ ⋆ *Suppose that $f : [a, b] \to \mathbb{R}$ is a function such that*

$$\lim_{x \to p} f(x)$$

exists for every $p \in [a, b]$. Prove that $f \in \mathscr{R}$ on $[a, b]$.

Proof. By the hypothesis, it means that both $f(p+)$ and $f(p-)$ exist and equal for every $p \in [a, b]$. Thus any discontinuity of f must be **simple** (see Definition 7.15 (Types of Discontinuity)). Therefore, by Theorem 7.16 (Countability of Simple Discontinuities), the set of all simple discontinuities of f is at most countable. Denote this set to be E. By the paragraph following Theorem 9.4 (The Lebesgue's Integrability Condition), E is of measure zero and hence we obtain our desired result, completing the proof of the problem. ∎

Problem 9.18

⋆ ⋆ *Suppose that $f : [0, 1] \to \mathbb{R}$ is continuous and there exists a positive constant M such that*

$$M \int_0^x f(t) \, \mathrm{d}t \leq f(x) \tag{9.51}$$

for all $x \in [0, 1]$. Prove that $f(x) \geq 0$ on $[0, 1]$.

Proof. For every $x \in [0, 1]$, we define

$$F(x) = \int_0^x f(t) \, \mathrm{d}t.$$

Since f is continuous on $[0,1]$, the First Fundamental Theorem of Calculus implies that

$$F'(x) = f(x) \geq MF(x) \tag{9.52}$$

on $[0,1]$. Next, we consider the function $G : [0,1] \to \mathbb{R}$ given by

$$G(x) = F(x)e^{-Mx}.$$

By the inequality (9.52), we deduce that

$$G'(x) = \frac{\mathrm{d}}{\mathrm{d}x}[F(x)e^{-Mx}] = e^{-Mx}F'(x) - Me^{-Mx}F(x) = e^{-Mx}[F'(x) - MF(x)] \geq 0$$

for every $x \in [0,1]$. By Theorem 8.9(a), G is monotonically increasing on $[0,1]$ and thus

$$G(x) \geq G(0) \tag{9.53}$$

for all $x \in [0,1]$. Since $G(0) = F(0)e^0 = 0$, we establish from the inequality (9.53) that

$$F(x)e^{-Mx} \geq 0$$

and then $F(x) \geq 0$ on $[0,1]$. Now we conclude from this and the hypothesis (9.51) that

$$f(x) \geq MF(x) \geq 0$$

on $[0,1]$. This completes the proof of the problem.   ∎

Problem 9.19

⋆ ⋆ *Suppose that $f : [0,1] \to \mathbb{R}$ is continuous on $[0,1]$. Let n be a positive integer. Prove that there exists a $\alpha \in [0,1]$ such that*

$$\int_0^1 x^n f(x)\,\mathrm{d}x = \frac{1}{n+1}f(\alpha). \tag{9.54}$$

Proof. Since f is continuous on $[0,1]$, the Extreme Value Theorem tells us that there exist $p, q \in [0,1]$ such that

$$f(p) = \sup_{x \in [0,1]} f(x) \quad \text{and} \quad f(q) = \inf_{x \in [0,1]} f(x).$$

By Theorem 9.6(a) and (b) (Operations of Integrable Functions), we see that

$$f(q)\int_0^1 x^n\,\mathrm{d}x \leq \int_0^1 x^n f(x)\,\mathrm{d}x \leq f(p)\int_0^1 x^n\,\mathrm{d}x. \tag{9.55}$$

By the Second Fundamental Theorem of Calculus, since

$$\frac{\mathrm{d}}{\mathrm{d}x}\left(\frac{x^{n+1}}{n+1}\right) = x^n,$$

we have

$$\int_0^1 x^n\,\mathrm{d}x = \frac{x^{n+1}}{n+1}\Big|_0^1 = \frac{1}{n+1}.$$

Thus we obtain from this and the inequalities (9.55) that

$$f(q) \le (n+1) \int_0^1 x^n f(x) \, \mathrm{d}x \le f(p).$$

Since f is continuous on $[0,1]$, we follow from the Intermediate Value Theorem that there exists a $\alpha \in [0,1]$ such that

$$(n+1) \int_0^1 x^n f(x) \, \mathrm{d}x = f(\alpha)$$

which implies the desired result (9.54). We end the proof of the problem. ∎

Problem 9.20

$(\star)(\star)$ *Suppose that $\psi : [a,b] \to \mathbb{R}$ has second derivative in $[a,b]$ and*

$$\psi(a) = \psi(b) = \psi'(a) = \psi'(b) = 0. \tag{9.56}$$

Prove that there exists a positive constant M such that

$$\left| \int_a^b \cos(\theta x) \psi(x) \, \mathrm{d}x \right| \le \frac{(b-a)M}{\theta^2}$$

for all $\theta > 1$.

Proof. Since ψ and $\cos(\theta x)$ are differentiable in $[a,b]$, we follow from the Integration by Parts that

$$\int_a^b \cos(\theta x) \psi(x) \, \mathrm{d}x = \int_a^b \psi(x) \frac{\mathrm{d}}{\mathrm{d}x} \left(\frac{\sin(\theta x)}{\theta} \right) \mathrm{d}x$$
$$= \frac{\sin(\theta x)}{\theta} \psi(x) \Big|_a^b - \frac{1}{\theta} \int_a^b \sin(\theta x) \psi'(x) \, \mathrm{d}x. \tag{9.57}$$

By the hypotheses (9.56), the expression (9.57) reduces to

$$\int_a^b \cos(\theta x) \psi(x) \, \mathrm{d}x = -\frac{1}{\theta} \int_a^b \sin(\theta x) \psi'(x) \, \mathrm{d}x. \tag{9.58}$$

Now we apply the Integration by Parts to the right-hand side of the expression (9.58) and then using the hypotheses (9.56) again, we derive that

$$-\frac{1}{\theta} \int_a^b \sin(\theta x) \psi'(x) \, \mathrm{d}x = \frac{1}{\theta} \int_a^b \psi'(x) \left(\frac{\cos(\theta x)}{\theta} \right)' \mathrm{d}x$$
$$= -\frac{1}{\theta^2} \int_a^b \cos(\theta x) \psi''(x) \, \mathrm{d}x. \tag{9.59}$$

Combining the two expressions (9.58) (9.59) and then using Theorem 9.6(f) (Operations of Integrable Functions), we obtain

$$\left| \int_a^b \cos(\theta x) \psi(x) \, \mathrm{d}x \right| = \left| \frac{1}{\theta^2} \int_a^b \cos(\theta x) \psi''(x) \, \mathrm{d}x \right| \le \frac{1}{\theta^2} \int_a^b |\cos(\theta x)| |\psi''(x)| \, \mathrm{d}x. \tag{9.60}$$

Since ψ'' is continuous on $[a, b]$, the Extreme Value Theorem ensures that there exists a positive constant M such that $|\psi''(x)| \le M$ for all $x \in [a, b]$. Furthermore, since $|\cos(\theta x)| \le 1$ for every $x \in [a, b]$ and $\theta > 1$, the inequality (9.60) can be reduced to

$$\left| \int_a^b \cos(\theta x)\psi(x)\,dx \right| \le \frac{(b-a)M}{\theta^2}$$

which is our desired result, completing the proof of the problem. ∎

Problem 9.21

(⋆) *Suppose that $f : [0, 1] \to [0, +\infty)$ is differentiable in $[0, 1]$ and $|f'(x)| \le M$ for some positive constant M. Let*

$$F(x) = \int_0^{f(x)} e^{-2t}\,dt,$$

where $x \ge 0$. Prove that

$$|F'(x)| \le M.$$

Proof. Since f is differentiable in $[0, 1]$ and $f(x) \ge 0$ on $[0, +\infty)$, it follows from the Chain Rule and the First Fundamental Theorem of Calculus that

$$F'(x) = \frac{d(f(x))}{dx} \cdot \frac{d}{d(f(x))} \int_0^{f(x)} e^{-2t}\,dt = f'(x)e^{-2f(x)}$$

so that

$$|F'(x)| \le \frac{|f'(x)|}{e^0} \le M$$

which is exactly the desired result. This completes the proof of the problem. ∎

Problem 9.22

(⋆)(⋆) *Suppose that $f : [a, b] \to \mathbb{R}$ has nth continuous derivative in $[a, b]$. Then we have*

$$f(x) = \sum_{k=0}^{n-1} \frac{f^{(k)}(p)}{k!}(x-p)^k + \frac{1}{(n-1)!} \int_p^x f^{(n)}(t)(x-t)^{n-1}\,dt, \qquad (9.61)$$

where $[p, x] \subseteq [a, b]$.

Proof. Since f' is continuous on $[a, b]$, $f' \in \mathscr{R}$ on $[a, b]$ by Theorem 9.3(a) and we follow from the Second Fundamental Theorem of Calculus that

$$f(x) - f(p) = \int_p^x f'(t)\,dt = -\int_p^x \underbrace{f'(t)}_{F(t)} \underbrace{\frac{d}{dt}(x-t)}_{G(t)}\,dt. \qquad (9.62)$$

Apply the Integration by Parts to the formula (9.62), we have

$$f(x) = f(p) - f'(t)(x-t)\Big|_p^x + \int_p^x f''(t)(x-t)\,dt$$

$$= f(p) + f'(p)(x - p) + \int_p^x f''(t)(x - t) \, dt. \tag{9.63}$$

Now we may express the formula (9.63) in the following form

$$f(x) = f(p) + f'(p)(x - p) - \frac{1}{2} \int_p^x \underbrace{f''(t)}_{F(t)} \underbrace{\frac{d}{dt}(x - t)^2}_{G(t)} \, dt$$

so that the Integration by Parts can be used again to obtain

$$f(x) = f(p) + f'(p)(x - p) + \frac{1}{2}f''(p)(x - p)^2 + \frac{1}{2} \int_p^x f'''(t)(x - t)^2 \, dt.$$

Since f has continuous derivative in $[a, b]$ up to order nth, the above process can be continued $(n - 3)$-steps to obtain

$$f(x) = \sum_{k=0}^{n-1} \frac{f^{(k)}(p)}{k!}(x - p)^k + \frac{1}{(n-1)!} \int_p^x f^{(n)}(t)(x - t)^{n-1} \, dt$$

which is our expected result (9.61).   ∎

Remark 9.7

Problem 9.22 says that we can express the remainder of Taylor's Theorem in an **exact form**.

Problem 9.23

(⋆) Prove the First Mean Value Theorem for Integrals by using the First Fundamental Theorem of Calculus.

Proof. Define

$$F(x) = \int_a^x f(t) \, dt. \tag{9.64}$$

Since f is continuous on $[a, b]$, we have F is differentiable in $[a, b]$ by the First Fundamental Theorem of Calculus. By the Mean Value Theorem for Derivatives, there exists a $p \in (a, b)$ such that

$$F(b) - F(a) = F'(p)(b - a). \tag{9.65}$$

By the definition (9.64), we have

$$F(b) = \int_a^b f(t) \, dt \quad \text{and} \quad F(a) = \int_a^b f(t) \, dt = 0$$

so that

$$F(b) - F(a) = \int_a^b f(t) \, dt. \tag{9.66}$$

Thus, by substituting the result (9.66) and the fact $F'(p) = f(p)$ into the formula (9.65), we have

$$\int_a^b f(t) \, dt = f(p)(b - a)$$

which is our desired result, completing the proof of the problem.   ∎

Problem 9.24

$\star$ $\star$ *Suppose that* $\varphi : [A, B] \to [a, b]$ *is a strictly increasing continuous onto function. Furthermore, suppose that* φ *has a continuous derivative on* $[A, B]$ *and* $f : [a, b] \to \mathbb{R}$ *is continuous in* $[a, b]$. *Prove that*

$$\int_{\varphi(A)}^{\varphi(B)} f(x)\, \mathrm{d}x = \int_A^B f(\varphi(y))\varphi'(y)\, \mathrm{d}y. \tag{9.67}$$

Proof. Since f is continuous on $[a, b]$, we have $f \in \mathscr{R}(\alpha)$ on $[a, b]$. In the Change of Variables Theorem, if we take $\alpha(x) = x$ which is monotonically increasing on $[a, b]$, then we have $\beta = \varphi$ and

$$\int_a^b f(x)\, \mathrm{d}x = \int_A^B f(\varphi)\, \mathrm{d}\varphi. \tag{9.68}$$

Since φ' is continuous on $[A, B]$, it follows from Theorem 9.3(a) that $\varphi' \in \mathscr{R}$ on $[A, B]$. Thus the Substitution Theorem implies that

$$\int_A^B f(\varphi)\, \mathrm{d}\varphi = \int_A^B f(\varphi(y))\varphi'(y)\, \mathrm{d}y. \tag{9.69}$$

Hence our desired result (9.67) follows by combining the two expressions (9.68) and (9.69) and using the facts that $\varphi(A) = a$ and $\varphi(B) = b$. This completes the proof of the problem. ∎

Problem 9.25

$\star$ $\star$ *Suppose that* $f : [a, b] \to \mathbb{R}$ *is continuous on* $[a, b]$. *Let* $[p, q] \subset [a, b]$ *and* x *be a variable such that* $[p + x, q + x] \subset [a, b]$. *Prove that*

$$\frac{\mathrm{d}}{\mathrm{d}x} \int_p^q f(x + y)\, \mathrm{d}y = f(q + x) - f(p + x).$$

Proof. Fix x, since $f(x + y) \in [a, b]$ for every $y \in [p, q]$, we have $f(x + y) \in \mathscr{R}$ on $[p, q]$. Let $\varphi : [p, q] \to [p + x, q + x]$ be defined by

$$\varphi(y) = x + y.$$

When $y = p$, $\varphi(p) = p + x$; when $y = q$, $\varphi(q) = q + x$. By Problem 9.24, we have

$$\frac{\mathrm{d}}{\mathrm{d}x} \int_p^q f(x + y)\, \mathrm{d}y = \frac{\mathrm{d}}{\mathrm{d}x} \int_p^q f(\varphi(y))\varphi'(y)\, \mathrm{d}y$$

$$= \frac{\mathrm{d}}{\mathrm{d}x} \int_{\varphi(p)}^{\varphi(q)} f(t)\, \mathrm{d}t$$

$$= \frac{\mathrm{d}}{\mathrm{d}x} \int_{p+x}^{q+x} f(t)\, \mathrm{d}t$$

$$= \frac{\mathrm{d}}{\mathrm{d}x} \left[\int_{p+x}^a f(t)\, \mathrm{d}t + \int_a^{q+x} f(t)\, \mathrm{d}t \right]$$

$$= \frac{\mathrm{d}}{\mathrm{d}x}\Big[- \int_a^{p+x} f(t)\,\mathrm{d}t + \int_a^{q+x} f(t)\,\mathrm{d}t \Big]. \tag{9.70}$$

Next, we apply the Chain Rule and then the First Fundamental Theorem of Calculus to the expression (9.70), we obtain

$$\frac{\mathrm{d}}{\mathrm{d}x} \int_p^q f(x+y)\,\mathrm{d}y = -f(p+x) + f(q+x) = f(q+x) - f(p+x),$$

as desired. We have completed the proof of the problem. ∎

Problem 9.26

⋆ ⋆ *Suppose $f : [a, b] \to \mathbb{R}$ is continuous on $[a, b]$. Prove that the formula*

$$2 \int_a^b \int_a^x f(x)f(y)\,\mathrm{d}y\,\mathrm{d}x = \Big(\int_a^b f(x)\,\mathrm{d}x \Big)^2 \tag{9.71}$$

holds.

Proof. For $x \in [a, b]$, let

$$F(x) = \int_a^x f(y)\,\mathrm{d}y.$$

Then we have

$$\int_a^b \int_a^x f(x)f(y)\,\mathrm{d}y\,\mathrm{d}x = \int_a^b f(x)\Big[\int_a^x f(y)\,\mathrm{d}y \Big]\,\mathrm{d}x = \int_a^b f(x)F(x)\,\mathrm{d}x. \tag{9.72}$$

Since f is continuous on $[a, b]$, the First Fundamental Theorem of Calculus implies that

$$F'(x) = f(x)$$

on $[a, b]$. Thus the expression (9.72) can be further rewritten as

$$\int_a^b \int_a^x f(x)f(y)\,\mathrm{d}y\,\mathrm{d}x = \int_a^b F(x)F'(x)\,\mathrm{d}x. \tag{9.73}$$

Apply the Integration by Parts to the right-hand side of the expression (9.73), we get

$$\int_a^b F(x)F'(x)\,\mathrm{d}x = F(b)F(b) - F(a)F(a) - \int_a^b F'(x)F(x)\,\mathrm{d}x$$

which means that

$$\int_a^b F(x)F'(x)\,\mathrm{d}x = \frac{1}{2}[F^2(b) - F^2(a)].$$

Since

$$F(b) = \int_a^b f(x)\,\mathrm{d}x \quad \text{and} \quad F(a) = 0,$$

we have

$$\int_a^b F(x)F'(x)\,\mathrm{d}x = \frac{1}{2}\Big(\int_a^b f(x)\,\mathrm{d}x \Big)^2. \tag{9.74}$$

Hence our desired result (9.71) follows immediately if we substitute the expression (9.74) back into the expression (9.73). This ends the proof of the problem. ∎

9.4 The Mean Value Theorems for Integrals

Problem 9.27

(⋆) Let $a \in \mathbb{R}$. Let $f : [a, a+1] \to \mathbb{R}$ be continuous and

$$\int_a^{a+1} f(x) \, \mathrm{d}x = 1.$$

Prove that $f(p) = 1$ for some $p \in (a, a+1)$.

Proof. A direct application of the First Mean Value Theorem for Integrals shows that there exists a $p \in (a, a+1)$ such that

$$1 = \int_a^{a+1} f(x) \, \mathrm{d}x = f(p)(a + 1 - a) = f(p).$$

This completes the proof of the problem. ∎

Problem 9.28

(⋆) Let $\alpha > 0$ and $n \in \mathbb{N}$, prove that

$$\int_n^{n+\alpha} \frac{\cos x}{x^2} \, \mathrm{d}x = \frac{\alpha \cos p}{p^2}$$

for some $p \in (n, n + \alpha)$.

Proof. Let $f : [n, n + \alpha] \to \mathbb{R}$ be defined by

$$f(x) = \frac{\cos x}{x^2}.$$

Since $n \geq 1$, it is clear that f is continuous on $[n, n + \alpha]$. By the First Mean Value Theorem for Integrals, *there exists* a $p \in (n, n + \alpha)$ such that

$$\int_n^{n+\alpha} \frac{\cos x}{x^2} \, \mathrm{d}x = \frac{\cos p}{p^2}(n + \alpha - n) = \frac{\alpha \cos p}{p^2}.$$

This completes the proof of the problem. ∎

Problem 9.29

(⋆)(⋆) Suppose that $\varphi : [a, b] \to \mathbb{R}$ is differentiable in $[a, b]$, $\varphi'(x) \geq \eta > 0$, φ' is monotonically decreasing and continuous on $[a, b]$. Prove that

$$\left| \int_a^b \sin \varphi(x) \, \mathrm{d}x \right| \leq \frac{2}{\eta}.$$

Proof. Since $\varphi'(x) > 0$ and φ' is monotonically decreasing on $[a, b]$, $\frac{1}{\varphi'}$ is well-defined, monotonically increasing on $[a, b]$ and

$$0 < \frac{1}{\varphi'(x)} \le \frac{1}{\eta}$$

on $[a, b]$. Since φ' is continuous on $[a, b]$, the function $\varphi' \sin \varphi$ is also continuous on $[a, b]$. Thus by the special case of the Second Mean Value Theorem For Integrals, we can find a $p \in [a, b]$ such that

$$\int_a^b \sin \varphi(x) \, dx = \int_a^b \underbrace{\frac{1}{\varphi'(x)}}_{f(x)} \times \underbrace{\varphi'(x) \sin \varphi(x)}_{g(x)} \, dx$$

$$= \frac{1}{\eta} \int_p^b [\sin \varphi(x)] \times \varphi'(x) \, dx. \tag{9.75}$$

Since $\varphi'(x) > 0$ on $[a, b]$, recall from Theorem 8.9(a) and Remark 8.3 that φ is strictly increasing on $[a, b]$. Since φ' is continuous on $[a, b]$, φ satisfies the conditions of Problem 9.24. It is clear that $f(x) = \sin x$ is continuous on $[\varphi(p), \varphi(b)]$, so we obtain from the formula (9.67) that

$$\int_p^b [\sin \varphi(x)] \times \varphi'(x) \, dx = \int_{\varphi(p)}^{\varphi(b)} \sin x \, dx = -\cos x \Big|_{\varphi(p)}^{\varphi(b)} = \cos \varphi(p) - \cos \varphi(b). \tag{9.76}$$

Hence we derive from the expressions (9.75) and (9.76) that

$$\left| \int_a^b \sin \varphi(x) \, dx \right| = \left| \frac{1}{\eta} [\cos \varphi(p) - \cos \varphi(b)] \right| \le \frac{2}{\eta},$$

completing the proof of the problem. ∎

Problem 9.30

(★)(★) Suppose that $f, g : [0, 1] \to \mathbb{R}$ are *monotonically increasing continuous functions.* Prove that

$$\int_0^1 f(x) \, dx \times \int_0^1 g(x) \, dx \le \int_0^1 f(x)g(x) \, dx.$$

Proof. By Theorem 9.6(e) (Operations of Integrable Functions), we see that $fg \in \mathcal{R}$ on $[0, 1]$. Define the function $\varphi : [0, 1] \to \mathbb{R}$ by

$$\varphi(x) = g(x) - \int_0^1 g(t) \, dt.$$

By the First Mean Value Theorem for Integrals, there exists a $p \in (0, 1)$ such that

$$\int_0^1 g(t) \, dt = g(p)$$

which gives

$$\varphi(x) = g(x) - g(p). \tag{9.77}$$

By Theorem 9.6 (Operations of Integrable Functions) again, we have $\varphi \in \mathscr{R}$ on $[0,1]$ which implies that $f\varphi \in \mathscr{R}$ on $[0,1]$. Furthermore, since g is monotonically increasing on $[0,1]$, we deduce from the result (9.77) that if $0 \le x \le p$, then $\varphi(x) \le 0$ and if $p \le x \le 1$, then $\varphi(x) \ge 0$. By these and the fact that f is monotonically increasing on $[0,1]$, we conclude that

$$
\int_0^1 f(x)\varphi(x)\,dx = \int_0^p f(x)\varphi(x)\,dx + \int_p^1 f(x)\varphi(x)\,dx
$$
$$
\ge f(p)\int_0^p \varphi(x)\,dx + f(p)\int_p^1 \varphi(x)\,dx
$$
$$
= f(p)\int_0^1 \varphi(x)\,dx. \tag{9.78}
$$

By the definition of φ, we must have

$$
\int_0^1 \varphi(x)\,dx = \int_0^1 g(x)\,dx - \int_0^1 \left(\int_0^1 g(t)\,dt\right) dx = 0 \tag{9.79}
$$

Therefore, we know from the inequality (9.78) and the result (9.79) that

$$
\int_0^1 f(x)\varphi(x)\,dx \ge 0
$$

and this implies that

$$
\int_0^1 f(x)g(x)\,dx \ge \int_0^1 f(x)\,dx \times \int_0^1 g(x)\,dx.
$$

This completes the proof of the problem.   ■

Problem 9.31

(⋆) Suppose that $f : [0, 2\pi] \to \mathbb{R}$ is strictly increasing on $[0, 2\pi]$ and $f(x) \ge 0$ on $[0, 2\pi]$. Prove that

$$
\int_0^{2\pi} f(x)\sin x\,dx < 0.
$$

Proof. It is clear that the function f satisfies the hypotheses in the special case of the Second Mean Value Theorem for Integrals, so there exists a $p \in (0, 2\pi)$ such that

$$
\int_0^{2\pi} f(x)\sin x\,dx = f(2\pi-)\int_p^{2\pi} \sin x\,dx
$$
$$
= f(2\pi-)\left[-\cos x\Big|_p^{2\pi}\right]
$$
$$
= -f(2\pi-)(1 - \cos p). \tag{9.80}
$$

Since f is strictly increasing on $[0, 2\pi]$ and $f(0) \ge 0$, we must have $f(2\pi-) > 0$. Since $p \in (0, 2\pi)$, we have $1 - \cos p > 0$. Hence these two facts imply that the right-hand side of the expression (9.80) must be negative. We end the proof of the problem.   ■

Problem 9.32

$(\star)(\star)(\star)$ Suppose that $f, g : [a, b] \to \mathbb{R}$ has continuous derivative on $[a, b]$. Suppose, further, that g is convex on $[a, b]$, $f(a) = g(a)$, $f(b) = g(b)$ and $g(x) \geq f(x)$ on $[a, b]$. Prove that

$$\int_a^b \sqrt{1 + [g'(x)]^2}\, dx \leq \int_a^b \sqrt{1 + [f'(x)]^2}\, dx.$$

Proof. Define the function $\varphi : \mathbb{R} \to \mathbb{R}$ by

$$\varphi(t) = \sqrt{1 + t^2}.$$

Since $\varphi''(t) = \frac{1}{(1+t^2)^{\frac{3}{2}}} \geq 0$ for all $t \in \mathbb{R}$, φ is convex on $\mathbb{R}$ by Theorem 8.16. Let $a < s < u < t < b$, if $\lambda = \frac{t-u}{t-s}$, then we have $0 < \lambda < 1$ and $\lambda s + (1 - \lambda)t = u$. Since φ is convex on (a, b), we have

$$\varphi(\lambda s + (1 - \lambda)t) \leq \lambda \varphi(s) + (1 - \lambda)\varphi(t)$$
$$\varphi(u) \leq \frac{t - u}{t - s}\varphi(s) + \left(1 - \frac{t - u}{t - s}\right)\varphi(t)$$
$$(t - s)\varphi(u) \leq (t - u)\varphi(s) + (u - s)\varphi(t)$$
$$\frac{\varphi(u) - \varphi(s)}{u - s} \leq \frac{\varphi(t) - \varphi(s)}{t - s}. \tag{9.81}$$

If we take $u \to s$ to both sides of the inequality (9.81), then we have

$$\varphi'(s) \leq \frac{\varphi(t) - \varphi(s)}{t - s}. \tag{9.82}$$

Since $f(x) \leq g(x)$ on $[a, b]$, we have $f'(x) \leq g'(x)$ on $[a, b]$. If we put $t = f'(x)$ and $s = g'(x)$ into the inequality (9.82), then we see that

$$\varphi(g'(x)) + [f'(x) - g'(x)]\varphi'(g'(x)) \leq \varphi(f'(x)). \tag{9.83}$$

Since f', g' and φ, φ' are continuous on $[a, b]$ and $\mathbb{R}$ respectively, they are Riemann integrable on $[a, b]$ and $\mathbb{R}$ respectively. By Theorem 9.5 (Composition Theorem), $\varphi(f'), \varphi(g')$ and $\varphi'(g')$ are Riemann integrable on $[a, b]$. Therefore, we apply Theorem 9.6(b) (Operations of Integrable Functions) to the inequality (9.83) to get

$$\int_a^b \varphi(g'(x))\, dx + \int_a^b [f'(x) - g'(x)]\varphi'(g'(x))\, dx \leq \int_a^b \varphi(f'(x))\, dx$$

which is equivalent to

$$\int_a^b \sqrt{1 + [g'(x)]^2}\, dx + \int_a^b [f'(x) - g'(x)]\varphi'(g'(x))\, dx \leq \int_a^b \sqrt{1 + [f'(x)]^2}\, dx. \tag{9.84}$$

Since g and φ are convex on (a, b) and $\mathbb{R}$ respectively, we follow from Theorem 8.15 that g' and φ' are monotonically increasing on (a, b) and $\mathbb{R}$ respectively. Thus $\varphi'(g')$ is also monotonically increasing on (a, b). In addition, since g' and φ' are continuous on $[a, b]$ and $\mathbb{R}$ respectively, the function $\varphi'(g')$ is continuous on $[a, b]$. Therefore, this implies that $\varphi'(g')$ is actually monotonically increasing on $[a, b]$. Hence it follows from the Second Mean Value Theorem for Integrals and then the Second Fundamental Theorem of Calculus that there exists a $p \in [a, b]$ such that

$$\int_a^b \underbrace{[f'(x) - g'(x)]}_{g(x)} \underbrace{\varphi'(g'(x))}_{f(x)} \, \mathrm{d}x = \varphi'(g'(a)) \int_a^p [f'(x) - g'(x)] \, \mathrm{d}x$$

$$+ \varphi'(g'(b)) \int_p^b [f'(x) - g'(x)] \, \mathrm{d}x$$

$$= \varphi'(g'(a))[f(p) - g(p) - f(a) + g(a)]$$

$$+ \varphi'(g'(b))[f(b) - f(p) - g(b) + g(p)]$$

$$= [\varphi'(g'(b)) - \varphi'(g'(a))][g(p) - f(p)]. \qquad (9.85)$$

Recall that $g(x) \geq f(x)$ on $[a, b]$ and $\varphi'(g')$ is monotonically increasing on $[a, b]$, we are able to obtain from the expression (9.85) that

$$\int_a^b [f'(x) - g'(x)]\varphi'(g'(x)) \, \mathrm{d}x \geq 0$$

so that the inequality (9.84) implies that

$$\int_a^b \sqrt{1 + [g'(x)]^2} \, \mathrm{d}x \leq \int_a^b \sqrt{1 + [f'(x)]^2} \, \mathrm{d}x.$$

This completes the proof of the problem.   ∎

APPENDIX A

Language of Mathematics

▌A.1 Fundamental Concepts

The goal of this appendix is to give a brief review to **mathematical logic** that we use frequently to write mathematical proofs logically and rigorously in this book. The main references we have used here are [1, chap. 9] and [12, §1.1 - §1.3].

▌A.1.1 What is logic?

This may be the first question in your mind. In fact, the term "logic" came from the Greek word "logos" which can be translated as "sentence", "reason", "rule" and etc. Of course, these translations are not enough to explain the specialized meaning of "logic" when one uses it nowadays.

Roughly speaking, logic is the study of **principles of correct and incorrect reasoning**. It is a tool to establish reasonable conclusions based on a given set of assumptions. A "logical" person wants to figure out what makes good/bad reasoning good/bad. Understanding such principles can keep us away from making mistakes in our own reasoning and they allow us to judge others' reasoning.

▌A.1.2 What is mathematical logic?

I think this is the second question in your mind. Briefly speaking, mathematical logic is a subfield of mathematics and it is the application of the theory and techniques of formal logic to mathematics and mathematical reasoning. One of the main characteristic features of mathematical logic is the use of mathematical language of symbols and formulas.

In advanced mathematics, you will be studying a lot of mathematical concepts and well-known results you have been already familiar with. It may happen that your *computational skills* are excellent. However, this is not what you are going to learn or sharpen in your undergraduate mathematics courses. Instead, we will emphasize the backbone of the theory behind. For examples,

- What is a rational number/irrational number/real number? How do we "count" those numbers?

- What is a continuous/differentiable/integrable function?

- Why do the Intermediate Value Theorem, the Mean Value Theorem for Derivatives, the Integration by Parts and other theorems that you used in a calculus course work?

In real analysis, we will answer the above questions in a systematical way so that you know not only *how* to apply such mathematical theorems, but also understand *why* they are true.

A.2 Statements and Logical Connectives

A.2.1 Statements

A **mathematical statement** or simply **statement** is a sentence which is either true or false, but *not both*. Usually, we apply the lowercase letters (e.g. p, q and r) to represent statements. When a statement is true, we assign its **truth value** to be truth, denoted by **T**; When the statement is false, its truth value is false and it is denoted by **F**. A **compound statement** is a statement known as the composition of statements by applying **logical connectives** such as "and", "or", "not", "if ... then" and "if and only if".

A.2.2 Logical Connectives

Let p and q be statements. The truth value of a compound statement of p and q can be expressed in terms of a **truth table**.

- **Conjunction.** The conjunction of statements p and q, denoted by $p \wedge q$ and read as "p and q". It is defined as true **only** when both p and q are true.

p	q	$p \wedge q$
T	T	T
T	F	F
F	T	F
F	F	F

Table A.1: The truth table of $p \wedge q$.

- **Disjunction.** The disjunction of statements p and q, denoted by $p \vee q$ and read as "p or q". It is defined as false **only** when both p and q are false.

p	q	$p \vee q$
T	T	T
T	F	T
F	T	T
F	F	F

Table A.2: The truth table of $p \vee q$.

- **Negation.** The negation of a statement p, denoted by $\sim p$ and read as "the negation of p". It is defined as the **opposite value** of p.

p	$\sim p$
T	**F**
F	**T**

Table A.3: The truth table of $\sim p$.

- **Conditional statement.** A conditional statement, symbolized by $p \to q$, is an "if-then" statement and it is read as "if p, then q". Here p is called the **hypothesis** and q is the **conclusion**. It is defined as false **only** when the hypothesis p is true and the conclusion q is false.

p	q	$p \to q$
T	**T**	**T**
T	**F**	**F**
F	**T**	**T**
F	**F**	**T**

Table A.4: The truth table of $p \to q$.

We notice that the **inverse** and the **converse** of the conditional statement $p \to q$ are

$$\sim p \to\sim q \quad \text{and} \quad q \to p$$

respectively.[a]

- **Biconditional statement.** A biconditional statement, symbolized by $p \leftrightarrow q$, is an "if and only if" statement and it is read as "p if and only if q". It is defined as true when *both* p and q have the same truth value.

p	q	$p \leftrightarrow q$
T	**T**	**T**
T	**F**	**F**
F	**T**	**F**
F	**F**	**T**

Table A.5: The truth table of $p \to q$.

A.2.3 Equivalent statements, Tautologies and Contradictions

- **Equivalent statements.** If two statements p and q have the same truth table, then we say that they are **equivalent**. For example, the statement $p \to q$ is equivalent to the statement $\sim q \to\sim p$. Symbolically, we write

$$p \to q \equiv\sim q \to\sim p.$$

---

[a]It can be seen that the inverse and the converse are **equivalent** (see §A.2.3) and furthermore, the inverse is the **contrapositive** (see §A.5.2) of the converse.

p	q	$\sim q$	$\sim p$	$p \to q$	$\sim q \to \sim p$
T	T	F	F	T	T
T	F	T	F	F	F
F	T	F	T	T	T
F	F	T	T	T	T

Table A.6: The truth table of $p \to q$ and $\sim q \to \sim p$.

- **Tautologies.** If a compound statement always takes the truth value "**T**" no matter what the truth values of the variables are, then we call such a compound statement a **tautology**.

p	q	$\sim q$	$\sim p$	$p \vee q$	$p \wedge q$	$(p \vee q) \vee (\sim p)$
T	T	F	F	T	T	T
T	F	T	F	T	F	T
F	T	F	T	T	F	T
F	F	T	T	F	F	T

Table A.7: The tautology $(p \vee q) \vee (\sim p)$

Particularly, we write "$p \Rightarrow q$" if $p \to q$ is a tautology.[b] For example, we have

$$x > 2 \Rightarrow x^2 > 4.$$

Similarly, we write "$p \Leftrightarrow q$" if $p \leftrightarrow q$ is a tautology.

- **Contradictions.** If a compound statement always takes the truth value "**F**" no matter what the truth values of the variables are, then we call such a compound statement a **contradiction**.

p	q	$\sim q$	$\sim p$	$p \vee q$	$p \wedge q$	$(p \wedge q) \wedge (\sim p)$
T	T	F	F	T	T	F
T	F	T	F	T	F	F
F	T	F	T	T	F	F
F	F	T	T	F	F	F

Table A.8: The contradiction $(p \wedge q) \wedge (\sim p)$

A.3 Quantifiers and their Basic Properties

A.3.1 Existential quantifier and universal quantifier

There are two types of quantifiers: **existential quantifier** and **universal quantifier**.

- **Existential quantifier.** The expression "$\exists x\ P(x)$" means "there exists an x such that the property $P(x)$ holds" or "there is at least one x such that the property $P(x)$ holds". Here, the notation $\exists$ is called the **existential quantifier**, and $\exists x$ means that at least one element x.

---

[b]The notation "$\Rightarrow$" is read as "**implies**".

- **Universal quantifier.** The expression "$\forall x\ P(x)$" can be interpreted as "for each/for all/for every/for any x, the property $P(x)$ is true". Here, the notation $\forall$ is called the **universal quantifier** and $\forall x$ means all the elements x.

- **Order of quantifiers.** On the one hand, the positions of the **same type** of quantifiers can be interchanged without affecting the truth value. For examples,

$$\forall x\ \forall y\ P(x,y) \Leftrightarrow \forall y\ \forall x\ P(x,y) \quad \text{and} \quad \exists x\ \exists y\ P(x,y) \Leftrightarrow \exists y\ \exists x\ P(x,y).$$

On the other hand, we can't switch the positions of **different types** of quantifiers, i.e.,

$$\exists x\ \forall y\ P(x,y) \not\Leftrightarrow \forall y\ \exists x\ P(x,y).$$

A.3.2 Properties of quantifiers

In many theorems or proofs, you may see one of the following four statements

$$\text{"}\forall x\ \forall y\ P(x,y)\text{''}, \quad \text{"}\forall x\ \exists y\ P(x,y)\text{''}, \quad \text{"}\exists x\ \forall y\ P(x,y)\text{''} \quad \text{and} \quad \text{"}\exists x\ \exists y\ P(x,y)\text{''}.$$

Understanding their interpretations can help you to *figure out* what the theorem is saying or what you are going to prove. Now their explanations are shown as follows:

- $\forall x\ \forall y\ P(x,y)$: For all x and for all y, the property $P(x,y)$ holds.

- $\forall x\ \exists y\ P(x,y)$: For all x, there exists y such that the property $P(x,y)$ holds.

- $\exists x\ \forall y\ P(x,y)$: There exists x such that for all y, the property $P(x,y)$ holds.

- $\exists x\ \exists y\ P(x,y)$: There exist x and y such that the property $P(x,y)$ holds.

Besides, we sometimes need to apply negation to quantifiers in writing a mathematical proof. Thus we have to understand what they are. In fact, the negation of the existential quantifier is the universal quantifier and vice versa. Symbolically, we have

$$\sim \forall x\ P(x) \equiv \exists x\ \sim P(x) \quad \text{and} \quad \sim \exists x\ P(x) \equiv \forall x\ \sim P(x).$$

Example A.1

The negation of the (true) statement "for all $n \in \mathbb{Z}$, we have $n^2 \geq 0$" is the (false) statement "there exists $n \in \mathbb{Z}$ such that $n^2 < 0$".

A.4 Necessity and Sufficiency

In mathematical logic, necessity and sufficiency are terms applied to describe an implicational relationship between statements. To say that p is a **necessary condition** for q means that it is impossible to have q without p. In other words, the nonexistence of p *guarantees* the nonexistence of q.

Example A.2

(a) The statement "$x^2 > 16$" is a **necessary condition** for the statement "$x > 5$". (It is impossible to have "$x > 5$" without having "$x^2 > 16$".)

(b) Having four sides is a **necessary condition** for being a rectangle. (It is impossible to
    <u>Statement p</u>   <u>Statement q</u>
    have a rectangle without having four sides.)

To say that the statement p is a **sufficient condition** for the statement q is to say that the existence of p *guarantees* the existence of q. In other words, it is impossible to have p without q: If p exists, then q must exist.

Example A.3

(a) The statement "$x > 5$" is a **sufficient condition** for the statement "$x^2 > 16$". (If "$x > 5$" is valid, then "$x^2 > 16$" is also valid.)

(b) Being a rectangle is a **sufficient condition** for having four sides. (If "being a rectangle" is valid, then "having four sides" is also valid.)

For equivalent statements p and q, we say that p is a **necessary and sufficient condition** of q. For example, the statement "a is even" is a necessary and sufficient condition for the statement "$(a + 1)^2 + 1$ is even".

A.5 Techniques of Proofs

Basically, there are three common ways of presenting a proof for a mathematical statement. They are

- **direct proof**,
- **proof by contrapositive** and
- **proof by contradiction**.

A.5.1 Direct Proof

To prove "$p \Rightarrow q$", we start with the hypothesis p and we proceed to show the truth of the conclusion q.

Example A.4

Prove that the sum of two even integers is also an even integer.
             <u>hypothesis p</u>   <u>conlusion q</u>

Proof. Let $2m$ and $2n$ be two even integers. Since we have

$$2m + 2n = 2(m + n),$$

the sum must be an even integer.   ∎

A.5.2   Proof by Contrapositive

Sometimes, it is *hard* to give a direct proof. In this case, one may give an "indirect proof" called "proof by contrapositive". By definition, the contrapositive of the statement $p \to q$ is $\sim q \to \sim p$. We note that a statement and its contrapositive are actually equivalent, i.e.,

$$p \to q \equiv \sim q \to \sim p.$$

See Table A.6 for this. Thus, to prove "$p \Rightarrow q$", it is equivalent to suppose that q is false (i.e., $\sim q$) and then we prove that p is also false (i.e., $\sim p$).

Example A.5

Prove that if $\underbrace{x^2 \text{ is even}}_{\text{hypothesis } p}$, then $\underbrace{x \text{ is even}}_{\text{conlusion } q}$.

Proof. The contrapositive of the statement is "If x is not even, then x^2 is not even" and we prove this is true. Since x is not even, it is odd. Thus we have $x = 2n + 1$ for some $n \in \mathbb{Z}$. This fact implies that

$$x^2 = (2n + 1)^2 = 4n^2 + 4n + 1 = 2(2n^2 + 2n) + 1$$

which is an odd integer.   ∎

A.5.3   Proof by Contradiction

It is an indirect proof. The basic idea of "proof by contradiction" is to assume that the statement we want to prove is **false** and then we prove that this assumption leads to a **contradiction**: a statement p and its negation $\sim p$ **cannot** both be true. The usual way of presenting a proof by contradiction is given as follows:

- **Step 1:** Assume that the statement p was false, i.e., $\sim p$ was true.

- **Step 2:** $\sim p$ implies both q and $\sim q$ are true for some statement q.

- **Step 3:** Since q and $\sim q$ cannot be both true, the statement p is **true**.

Example A.6

Prove that $\underbrace{\text{for all } m, n \in \mathbb{Z}, \text{ we have } m^2 - 4n \neq 2}_{\text{statement } p}$.

Proof. The negation of p is that "there exists $m, n \in \mathbb{Z}$ such that $m^2 - 4n^2 = 2$". It is clear from $m^2 - 4n^2 = 2$ that

$$m^2 = 4n + 2 = 2(2n + 1)$$

so that m^2 is even. Since m^2 is even, m must also be even. Let $m = 2k$ for some $k \in \mathbb{Z}$. Put $m = 2k$ into the original equation $m^2 - 4n = 2$, we obtain

$$(2k)^2 - 4n = 2$$
$$2(k^2 - n) = 1. \tag{A.1}$$

Since $k^2 - n \neq 0$, the equation (A.1) shows that $\underbrace{1 \text{ is even}}_{\text{statement } q}$. However, it is well-known that $\underbrace{1 \text{ is } \textbf{not} \text{ even}}_{\text{negation } \sim q}$. This is obvious a contradiction. Hence the statement p must be true.   ∎

Index

Symbols
n-th partial sums, 75
nth derivative of f, 131

A
A.M. $\geq$ G.M., 71, 93, 168
Abel's Test, 78
absolute convergence, 77
absolute maximum, 129
absolute minimum, 129
absolute value, 9
algebraic, 24
alternating series, 77
antiderivative, 161
Archimedean Property, 10
at most countable, 19

B
biconditional statement, 189
bijective, 2
Bolzano-Weierstrass Theorem, 68, 138
Bonnet's Theorem, 162
bounded sequence, 49
bounded set, 30, 100

C
Cantor set, 45
Cantor's intersection theorem, 30
cardinality, 19
cartesian product, 2
Cauchy Criterion, 76
Cauchy Mean Value Theorem, 129
Cauchy product, 78
Cauchy sequence, 52
Chain Rule, 128
Change of Variables Theorem, 160
circle of convergence, 79

closed set, 27, 99
closure, 28
compact metric space, 30, 42, 52
compact set, 30, 100
comparability, 3
comparison test, 76
complement, 2
complete metric space, 52, 70
Completeness Axiom, 10, 100
composition, 2
compound statement, 188
concave functions, 132
conclusion, 189
conditional statement, 189
conjunction, 188
connected set, 31, 100
continuity of functions, 98
continuous at p, 98
continuous on E, 98
contradictions, 190
converge, 49
converse, 189
convex functions, 132
convexity, 132
countability, 19
countable, 19
cover, 29

D
Darboux's Theorem, 129
definite integrals, 157
derivative of f, 127
difference, 1
differentiable at x, 127
differentiable on E, 127
differentiation, 127
direct proof, 192

Dirichlet function, 104, 165
Dirichlet product, 78
Dirichlet's Test, 78
discontinuity of the first kind, 101
discontinuity of the second kind, 101
discontinuous at p, 101
disjunction, 188
distance, 9, 27
distance function, 27
diverge, 49
domain, 2
dummy variable, 158

E
empty set, 1
equivalence class, 3
equivalence relation, 3
equivalent, 189
equivalent statements, 189
existential quantifier, 190
Extended real number system, 51
Extreme Value Theorem, 100

F
Fermat's Theorem, 129
finite intersection property, 43
First Fundamental Theorem of Calculus,
    161
fixed point, 119
Froda's Theorem, 102
function, 2

G
geometric series, 77
graph of f, 132
greatest integer function, 121

H
harmonic series, 81
Heaviside step function, 160
Heine-Borel Theorem, 30, 100
hypothesis, 189

I
implies, 190
indeterminate form, 130
indirect proof, 193
infinite series, 75
injective, 2

integer, 1
Integration by Parts, 162
interior, 28
interior point, 28
Intermediate Value Theorem, 100, 129
Intermediate Value Theorem for
    Derivatives, 130
intersection, 1
inverse, 189
inverse function, 100
inverse map, 2
irrational number, 1

J
Jensen's inequality, 155

L
L'Hôspital's Rule, 130
L'Hospital's Rule, 130
Lebesgue's Integrability Condition, 159
left continuous at p, 120
left-hand derivative, 127
left-hand limit, 101
Leibniz's rule, 131
limit of functions, 97
limit point, 28
limits at infinity, 102
limits of sequence, 49
Lipschitz condition, 142
Lipschitz constant, 142
local extreme of f, 128
local maximum at p, 117
local minimum at p, 117
logical connectives, 188
lower limit, 51
lower Riemann integral, 158

M
mapping, 2
mathematical statement, 188
Mean Value Theorem for Derivatives, 129,
    162
measure zero, 159
metric, 27
metric space, 27
Monotonic Convergence Theorem, 50
monotonic functions, 102
monotonic sequence, 50
monotonically decreasing, 50, 102, 129

monotonically increasing, 50, 102, 129

N
necessary and sufficient condition, 192
necessary condition, 191
necessity, 191
negation, 189
neighborhood, 27
nonreflexivity, 3

O
one-to-one, 2
onto, 2
open cover, 30
open set, 27, 99
order relation, 3
ordered pairs, 2

P
partial summation formula, 78
partition, 3, 157
periodic, 116
points, 27
popcorn function, 167
positive integer, 1
power series, 79
power set, 20
primitive function, 161
proof by contradiction, 192
proof by contrapositive, 192
proper subset, 1

R
radius, 27
radius of convergence, 79
range, 2
ratio test, 76
rational nunbers, 1
real number, 1
real number line, 9
rearrangement of series, 79
refinement, 157
reflexivity, 3
relation, 3
Riemann function, 105, 167
Riemann Integrability Condition, 159
Riemann integral, 158
Riemann-Stieltjes integral, 158
right continuous at p, 120

right-hand derivative, 127
right-hand limit, 101, 120
root test, 76
rule of assignment, 2
ruler function, 167

S
Sandwich Theorem, 50
Schwarz inequality, 90
Schwarz Inequality for Integral, 167
Second Fundamental Theorem of Calculus, 161
separation, 31
series, 75
set, 1
sign-preserving Property, 112
simple discontinuity, 101
Squeeze Theorem for Convergent Sequences, 50, 98
statement, 188
strictly convex, 132
strictly decreasing, 102, 129
strictly increasing, 102, 129
subsequence, 50
subsequential limit, 51
subset, 1
Substitution Theorem, 160
sufficiency, 191
sufficient condition, 192
surjective, 2
symmetry, 3

T
tautologies, 190
Taylor's polynomial, 131
Taylor's Theorem, 131, 179
Thomae's function, 167
transitivity, 3
translation, 32
truth table, 188
truth value, 188

U
uncountable, 19
uniform continuity, 99
union, 1
unit step function, 160
universal quantifier, 191
upper limit, 51

upper Riemann integral, 158

V
value, 2

variable of integration, 158
vector space, 164

Bibliography

[1] Z. Adamowicz and P. Zbierski, *Logic of Mathematics: a Modern Course of Classical Logic*, John Wiley & Sons Inc., 1997.

[2] Tom M. Apostol, *Calculus Vol. 1*, 2nd ed., John Wiley & Sons, Inc., 1967.

[3] Tom M. Apostol, *Mathematical Analysis*, 2nd ed., Addison-Wesley Publishing Company, 1974.

[4] Asuman G. Aksoy and Mohamed A. Khamsi, *A Problem Book in Real Analysis*, Springer-Verlag, New York, 2009.

[5] Robert G. Bartle and Donald R. Sherbert, *Introduction to Real Analysis*, 4th ed., John Wiley & Sons, Inc., 2011.

[6] Patrick M. Fitzpatrick, *Advanced Calculus*, 2nd ed., Brooks/Cole, 2006.

[7] G. H. Hardy, *A Course of Pure Mathematics*, 10th ed., Cambridge University Press, 2002.

[8] Masayoshi Hata, *Problems and Solutions in Real Analysis*, Hackensack, NJ: World Scientific, 2007.

[9] J. Lüroth, Bemerkung über gleichmässige Stetigkeit, *Math. Ann.*, Vol. 6, 1873, pp. 319, 320.

[10] James R. Munkres, *Topology A First Course*, 2nd ed., Prentice-Hall Inc., 2000.

[11] Teodora-Liliana T. Rădulescu, Vincenţiu D. Rădulescu and Titu Andreescu, *Problems in Real Analysis: Advanced Calculus on the Real Axis*, Springer-Verlag, New York, 2009.

[12] W. Rautenberg, *A Concise Introduction to Mathematical Logic*, 3rd ed., Springer-Verlag, New York, 2010.

[13] W. Rudin, *Principles of Mathematical Analysis*, 3rd ed., Mc-Graw Hill Inc., 1976.

[14] Rami Shakarchi, *Problems and Solutions for Undergraduate Analysis*, Springer-Verlag, New York, 1998.

[15] Terence Tao, *Analysis I*, 3rd ed., Singapore: Springer, 2016.

[16] V. A. Zorich, *Mathematical Analysis I*, Springer-Verlag, New York, 2004.

Made in the USA
Middletown, DE
12 September 2019